Praise for *The Death of Distance*

'If you want to know what is going on now in the information
 communications technologies, and how they will impact the
 dustry and our society in the not so distant future, this book is
 ntial reading.'

Rupert Murdoch, Chairman and Chief Executive,
The News Corporation

' markets open, as technologies converge and as our industry
 es itself, everyone in telecommunications knows our only
ce ty is that nothing will ever be the same again. *The Death of
D: e* captures the fundamental – yet exhilarating – changes we
ar eriencing and provides an insight into the ever more profound
im communications will have on our lives in the future.'

Sir Iain Vallance, Chairman, British Telecommunications plc

' oss manages to distill into a single readable volume almost
 rrent thinking about the subject ... More important, she adds
  judiciously weighing the impact of individual innovations,
 ting those that will matter and the timescale over which
 pact will be felt ... Cairncross's underlying approach is a sort
 optimism [and] optimism is in short supply in futurology
 he choice is often between hysterical gloom and Panglossian
 logy-worship. In avoiding these extremes, Cairncross has per-
 l a genuine service.'

Peter Martin, *Financial Times*

'Cairncross uses her crystal ball, with the aid of bar-graphs and pie-
charts to predict the new world order. Hers is a libertarian and
cautiously utopian vision to balance against the Cassandra voices
o' loom-mongers.'

Steven Poole, *The Guardian*

To my parents

The Death of Distance

How the Communications Revolution Will Change Our Lives

Frances Cairncross
of *The Economist*

ORION
BUSINESS
BOOKS

The author hereby asserts her moral right to be identified as the Author of this work.

Hardback edition first published in Great Britain in 1997
This paperback edition first published in Great Britain in 1998 by
Orion Business
An imprint of The Orion Publishing Group Limited
Orion House, 5 Upper St Martin's Lane, London WC2H 9EA

First published in the United States by Harvard Business School Press, 1997.
This edition by arrangement with The Orion Publishing Group Limited.

A CIP catalogue record for this book is available in the British Library.

ISBN 0–75281–252–1

Printed and bound in Great Britain by
Butler & Tanner Ltd, Frome and London

Contents

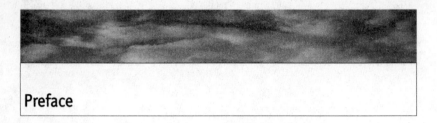

Preface

Describing the electronic miracles of our age in the old-fashioned format of ink on wood pulp may strike you as ironic. Put it down to the fact that, for the moment, the printed and bound book remains the most convenient way to introduce new ideas into the world. The new ideas in this book are about the many ways in which the most significant technological changes of our time will affect the next century— and your life. You will find a preview of the most important in "The Trendspotter's Guide to New Communications" that immediately follows this preface; the rest of the book sets out to interpret and elaborate these core points.

You need not be an Internet enthusiast to benefit from this book. The book is only incidentally about communications technologies themselves, and you certainly do not need to understand microprocessors, digital compression, fiber-optic cable, or any other technical paraphernalia. You need only a desire to know how these technologies will transform our lives and those of our children.

Forecasting the future is easier when the big technological innovations are in place. In 1897, for example, a far-sighted person could have guessed that the invention of the automobile would change society dramatically in the subsequent century. The design of the car and its engine have changed many times over the past hundred years, but the basic technology and its potential were already in place. The advances of the past few decades are now converging. Creators and users of technologies such as the Internet, the mobile telephone, and digital television will refine and rearrange them in many ways in the coming century, but their broad shape is clear to us.

. . . . .

The book began as a survey, also called "The Death of Distance," that I wrote for *The Economist* in 1995.[1] The title plays upon *The Tyranny of Distance*, Geoffrey Blainey's classic study of the impact of isolation on Australia.[2] After having various conversations with people at the World Bank and reading "Near-Zero Tariff Telecommunications,"[3] I realized that the steep fall in the distance premium for communications would be of enormous economic and social importance around the world.

Many people have contributed to the ideas in the book. Among those who have read and commented on various drafts are Patrick Barwise, David Bowen, Jonathan Davis, Jason Kowal, Patrick Lane, Sverker Lindbo, Edward Lucas, Robert Pinder, Nick Valery, Hal Vogel, and Pam Woodall. Victor Earl and Emma Whitehouse checked many facts for me, as did Tina Davis, Jenny Geddes, and Fiona Haynes. Michael Minges, Sam Paltridge, Greg Staple, and Mark Roberts answered innumerable questions.

I owe an enormous debt to my publishers, who have once again been an object lesson in what good editing should be. In particular, my editor Kirsten Sandberg, copyeditor Susan Boulanger, and managing editor Barbara Roth improved the original draft beyond all recognition—a service publishers rarely perform for authors these days. They have been stimulating and reassuring throughout the process.

I owe my greatest debts of all to four people: Chris Anderson, Azeem Azhar, Tim Kelly, and Hamish McRae. Chris and Azeem, my colleagues at *The Economist*, have been endlessly generous with their ideas, their understanding of communications, and their time. They have read drafts, offered facts and examples that galvanized the text, and rescued me from egregious errors. Tim Kelly of the International Telecommunication Union in Geneva also read the whole book, some of it more than once, and made numerous detailed and inestimable suggestions. And Hamish McRae not only put up with all the usual inconveniences that the spouses of authors endure, but also suggested many of the most illuminating and original points. Those who have read his own book, *The World in 2020*,[4] will spot a certain continuity of thought. This book is really a joint effort, the product of many conversations over breakfast, on walks across Hampstead Heath, and far into the night.

London, July 1997

Introduction to Paperback Edition

This book is the story of the world's third great transport revolution. One of its main themes is the way that technological change shapes the economy and society. Of all the technological changes of the past two centuries, few have had as big an impact on the way people live and think—and, indeed, on the natural world too—as those which have reduced the cost and increased the speed, capacity and reach of transport.

In the first revolution, which dominated the 19th century, it was the transport of goods that was mainly transformed. When the century opened, Thomas Malthus—who put the "dismal" into the science of economics—had just predicted that Europe would starve, because the growth of population would outstrip that of goods. Instead, the Victorians enjoyed a century when Europe's population grew faster than it had ever done before or since. The coming of the railway and the steamship, and the consequent decline in the cost of transporting goods from one continent to another, allowed Europeans to trade the output of their industries for American grain.

In this century, it is the transport of people that has been revolutionized. The car has brought the luxury of personal mobility to millions, allowing individuals to choose spouses, homes and jobs many miles from the villages where their parents spent their days. In its wake has also come a transformation in the appearance of cities and countryside, with the spread of suburbs, superstores and second homes. In the past half-century, the falling cost of air travel has brought a different kind of revolution in the mobility of people: it has knitted continents together, by carrying business people, tourists and immigrants to destinations that were once far beyond their reach.

The revolution that will shape the next century is in the transport of ideas and information. The fall in the cost and increase in capacity of communications over long distances herald a revolution of a different sort: self-reinforcing and self-refining. For not only can information travel round the world on an unprecedented scale; so can ideas on how to use that information. Innovation thus becomes a global game that everyone can play. If one of the great developments of the past century has been the invention of the process of invention, the death of distance will make that process faster, and its results more rapidly accessible.

In the year since this book was first published, the changes it describes have continued with astonishing speed. Long-distance and international telephone charges have continued their steep fall; the carrying capacity of the world's communications networks has multiplied; mobile telephones, already more than half the world's new subscriptions, have become increasingly versatile; the Internet, in more and more countries, has begun the transformation from a curiosity to a routine business tool.

At the same time, the sheer unpredictability of the revolution has also become more evident. For instance, many of the companies that first made a splash on the Internet have found it hard to turn hype into profit. More ominously, the vulnerability of a networked world has become more apparent as the millennium approaches, and companies worry about the effects of "bugs" in one part of a system on other, de-bugged parts.

Another revelation has been the importance of size. On the face of it, the falling costs of computing and communicating, lower many barriers to entry by giving individuals and small firms advantages that were once available only to large companies. A Web site (www.deathofdistance.com) set up by a friend of the author and linked to an Internet bookstore has sold several hundred copies of this book: as many as a large London bookshop might have moved in the same period. However, a networked world also has powerful forces for concentration. It encourages common standards—such as Microsoft's Windows; it gives economic advantages to big networks over small— see the concentration in America's telecoms industry; and it invests brands with a new power. There will be more pygmies in this new world, but there will also be more giants.

Of all the issues raised by the death of distance, one of the most significant is surely that of winners and losers. Many fear that the losers will be the poorest and weakest, and that the revolution will aggravate inequality, not diminish it.

One reply to such fears is that the changes ahead are likely to make the world overall wealthier, faster than would otherwise be the case. Ideas and information are the essential building blocks of growth; they will now spread faster and further. Farmers in developing countries will have new access to information on prices and crop technologies; doctors will learn faster of effective new treatments; entrepreneurs will be able to locate markets that would once have been out of reach. Overall, the gains of winners are likely to exceed the losses of losers. Among the biggest winners will be those developing countries that embrace new kinds of communications and so bring down the cost of a telephone or an Internet link. Give a lorry driver in such a country a mobile telephone, or a fisherman a satellite link, and you turbo-charge the whole process of development.

A second reassurance to those who worry about "techno ghettos" is that this revolution is powered by falling prices. The declining cost of communicating brings it within reach of those for whom it was once a luxury—the young, the housebound elderly with children far away, the poor who once could not afford a telephone. It offers them a way into the job market; access to education; contact with family and friends.

In the reception of the first edition of this book, the main criticism was that it was too optimistic. There may be justice in this; but those who seek a pessimistic analysis of the death of distance have plenty of pundits to choose from. This transport revolution will be at least as beneficial to human happiness as its two predecessors. Globalization has brought plenty of problems: environmental degradation; loss of jobs and unemployment for some; insecurity for others. But it has also brought an unprecedented rise in life expectancy, health and living standards. A smaller share of humanity is hungry, and a larger share literate, than ever before. These are prizes worth having. The death of distance will bring to many more people around the world the privileges and pleasures that up to now have been available only to the rich.

London, June 1998

How will the death of distance shape the future? Here are some of the most important developments to watch, each discussed in depth later in this book.

1. **The Death of Distance.** Distance will no longer determine the cost of communicating electronically. Companies will organize certain types of work in three shifts according to the world's three main time zones: the Americas, East Asia/Australia, and Europe.
2. **The Fate of Location.** No longer will location be key to most business decisions. Companies will locate any screen-based activity anywhere on earth, wherever they can find the best bargain of skills and productivity. Developing countries will increasingly perform on-line services—monitoring security screens, running help-lines and call centers, writing software, and so forth—and sell them to the rich industrial countries that generally produce such services domestically.
3. **The Irrelevance of Size.** Small companies will offer services that, in the past, only giants had the scale and scope to provide. Individuals with valuable ideas, initiative, and strong business plans will attract global venture capital and convert their ideas into viable businesses. Small countries will also be more viable. That will be good news for secession movements everywhere.
4. **Improved Connections.** Most people on earth will eventually have access to networks that are all switched, interactive, and broadband: "switched," like the telephone, and used to contact many other subscribers; "interactive" in that, unlike broadcast TV, all ends of the network can communicate; and "broad-

band," with the capacity to receive TV-quality motion pictures. While the Internet will continue to exist in its present form, it will also be integrated into other services, such as the telephone and television.

5. **More Customized Content.** Improved networks will also allow individuals to order "content for one"; that is, individual consumers will receive (or send) exactly what they want to receive (or send), when, and where they want it.

6. **A Deluge of Information.** Because people's capacity to absorb new information will not increase, they will need filters to sift, process, and edit it. Companies will have greater need of boosters—new techniques—to brand and push their information ahead of the competition's.

7. **Increased Value of Brand.** What's hot—whether a product, a personality, a sporting event, or the latest financial data—will attract greater rewards. The costs of producing or promoting these commodities will not change, but the potential market will increase greatly. That will create a category of global super-rich, many of them musicians, actors, artists, athletes, and investors. For the successful few and their intermediaries, entertaining will be the most lucrative individual activity on earth.

8. **Increased Value in Niches.** The power of the computer to search, identify, and classify people according to similar needs and tastes will create sustainable markets for many niche products. Niche players will increase, as will consumers' demand for customized goods and services.

9. **Communities of Practice.** The horizontal bonds among people performing the same job or speaking the same language in different parts of the world will strengthen. Common interests, experiences, and pursuits rather than proximity will bind these communities together.

10. **Near-Frictionless Markets.** Many more companies and customers will have access to accurate price information. That will curtail excessive profits, enhance competition, and help to curb inflation, resulting in "profitless prosperity": it will be easier to find buyers, but hard to make fat margins.

11. **Increased Mobility.** Every form of communication will be available for mobile or remote use. While fixed connections such as

cable will offer greater capacity and speed, wireless will be used, not just to send a signal over a large region, but to carry it from a fixed point to users in a relatively small radius. Satellite transmission will allow people to use a single mobile telephone anywhere, and the distinctions between fixed and mobile receiving equipment (a telephone or a personal computer) will blur.

12. **More Global Reach, More Local Provision.** While small companies find it easier to reach markets around the world, big companies will more readily offer high-quality local services, such as putting customers in one part of the world directly in touch with expertise in other places, and monitoring more precisely the quality of local provision.

13. **The Loose-Knit Corporation.** Culture and communications networks, rather than rigid management structures, will hold companies together. Many companies will become networks of independent specialists; more employees will therefore work in smaller units or alone. Loyalty, trust, and open communications will reshape the nature of customer and supplier contracts: suppliers will draw directly on information held in databases by their customers, working as closely and seamlessly as an in-house supplier now does. Technologies such as electronic mail and computerized billing will reduce the costs of dealing with consumers and suppliers at arm's length.

14. **More Minnows, More Giants.** On one hand, the cost of starting new businesses will decline, and companies will more easily buy in services so that more small companies will spring up. On the other, communication amplifies the strength of brands and the power of networks. In industries where networks matter, concentration may increase, but often in the form of loose global associations under a banner of brands or quality guarantees.

15. **Manufacturers as Service Providers.** Feeding information on a particular buyer's tastes straight back to the manufacturer will be easier and so manufacturers will design more products specially for an individual's requirements. Some manufacturers will even retain lasting links with their products: car companies, for instance, will continue electronically to track, monitor, and learn about their vehicles throughout the product life cycle. New

· · · · ·

opportunities to provide services for customers will emerge, and
some manufacturers may accept more responsibility for dispos-
ing of their products at the end of the cycle.

16. **The Inversion of Home and Office.** As more people work from
home or from small, purpose-built offices, the line between work
and home life will blur. The office will become a place for the
social aspects of work such as celebrating, networking, lunching,
and gossiping. Home design will also change, and the domestic
office will become a regular part of the house.

17. **The Proliferation of Ideas.** New ideas and information will travel
faster to the remotest corners of the world. Third world countries
will have access to knowledge that the industrial world has long
enjoyed. Communities of practice and long-distance education
programs will help people to find mentors and acquire new skills.

18. **A New Trust.** Since it will be easier to check whether people and
companies deliver what they have promised, many services will
become more reliable and people will be more likely to trust each
other to keep their word. However, those who fail to deliver will
quickly lose that trust, which will become more difficult to regain.

19. **People as the Ultimate Scarce Resource.** The key challenge for
companies will be to hire and retain good people, extracting
value from them, rather than allowing them to keep all the value
they create for themselves. A company will constantly need to
convince its best employees that working for it enhances each
individual's value.

20. **The Shift from Government Policing to Self-Policing.** Governments will find national legislation and censorship inadequate
for regulating the global flow of information. As content sweeps
across national borders, it will be harder to enforce laws banning
child pornography, libel, and other criminal or subversive material and those protecting copyright and other intellectual property.
But greater electronic access to information will give people better means to protect themselves. The result will be more individual responsibility and less government intervention.

21. **Loss of Privacy.** As in the village of past centuries, protecting privacy will be difficult. Governments and companies will easily
monitor people's movements. Machines will recognize physical
attributes like a voice or fingerprint. People will thus come to

embody their identity. Civil libertarians will worry, but others will accept the loss as a fair exchange for the reduction of crime, including fraud and illegal immigration. In the electronic village, there will be little true privacy—and little unsolved crime.

22. **Redistribution of Wages.** Low-wage competition will reduce the earning power of many people in rich countries employed in routine screen-based tasks, but the premium for certain skills will grow. People with skills that are in demand will earn broadly similar amounts wherever they live in the world. So income differences within countries will grow; and income differences between countries will narrow.

23. **Less Need for Immigration and Emigration.** Poor countries with good communications technology will be able to retain their skilled workers, who will be less likely to emigrate to countries with higher costs of living if they can earn rich-world wages and pay poor-world prices for everyday necessities right at home. Thus inexpensive communications may reduce some of the pressure to emigrate.

24. **A Market for Citizens.** The greater freedom to locate anywhere and earn a living will hinder taxation. Savers will be able to compare global investment rates and easily shift money abroad. High-income earners and profitable companies will be able to move away from hefty government-imposed taxes. Countries will compete to bid down tax rates and to attract businesses, savers, and wealthy residents.

25. **Rebirth of Cities.** As individuals spend less time in the office and more time working from home or traveling, cities will transform from concentrations of office employment to centers of entertainment and culture; that is, cities will become places where people go to stay in hotels, visit museums and galleries, dine in restaurants, participate in civic events, and attend live performances of all kinds. In contrast, some poor countries will stem the flight from the countryside to cities by using low-cost communications to provide rural dwellers with better medical services, jobs, education, and entertainment.

26. **The Rise of English.** The global role of English as a second language will strengthen as it becomes the common standard for telecommunicating in business and commerce. Many more

countries, especially in the developing world, will therefore adopt English as a subsidiary language. It will be as important to learn English as to use software that is compatible with the near-universal MS-DOS.

27. **Communities of Culture.** At the same time, electronic communications will reinforce less widespread languages and cultures, not replace them with Anglo-Saxon and Hollywood. The declining cost of creating and distributing many entertainment products and the corresponding increase in production capacity will also reinforce local cultures and help scattered peoples and families to preserve their cultural heritage.

28. **Improved Writing and Reading Skills.** Electronic mail will induce young people to express themselves effectively in writing and to admire clear and lively written prose. Dull or muddled communicators will fall by the information wayside.

29. **Rebalance of Political Power.** Since people will communicate their views on government more directly, rulers and representatives will become more sensitive (and, perhaps, more responsive) to lobbying and public-opinion polls, especially in established democracies. People who live under dictatorial regimes will make contact more easily with the rest of the world.

30. **Global Peace.** As countries become even more economically interdependent and as global trade and foreign investment grow, people will communicate more freely and learn more about the ideas and aspirations of human beings in other parts of the globe. The effect will be to increase understanding, foster tolerance, and ultimately promote worldwide peace.

Now read on.

1
The Communications Revolution

If the miles separate you from those you love, take heart. On Mother's Day, when Americans make more long-distance calls than on any other day of the year, MCI, the country's second-largest long-distance telephone company, likes to give its regular customers a treat: in 1995 and again in 1996, their calls to one another were free. In time, every day will be Mother's Day, everywhere. It will be no more expensive to telephone someone on the other side of the world than to talk to someone in the house across the street. In fact, it will be just like having your mother next door.

The death of distance as a determinant of the cost of communicating will probably be the single most important force shaping society in the first half of the next century. Technological change has the power to revolutionize the way people live, and this one will be no exception. It will alter, in ways that are only dimly imaginable, decisions about where people work and what kind of work they do, concepts of national borders and sovereignty, and patterns of international trade. Its effects may well be as pervasive as those of the discovery of electricity, which led in time to the creation of the sky-scraper cities of Manhattan and Hong Kong and transformed labor productivity in the home.

But the death of distance is only one of the astonishing changes taking place as communications and computers are combined in new ways. Fiber-optic networks and digital compression will allow many families, sometime in the first decade of the next century, to receive a personalized "television channel" that makes available tens of thousands of films and programs. Networks are being developed that are (a) like the telephone, "switched" so that they can be used by many subscribers; (b) like television, high capacity, or "broadband" so that they can carry moving pictures; and (c) interactive, so that, unlike with television, every user of the network can communicate with every other part. Mobile telephones now account for almost half of all new telephone connections worldwide. Tracking systems are now so refined that they allow companies with large truck fleets to monitor whether their drivers waste fuel by driving in the wrong gear. The Internet, virtually unheard of a decade ago, was being used in early 1997 by an estimated fifty-seven million people around the world, with perhaps another thirteen million or so using it just for electronic mail.[1]

That these technologies will change the world is beyond a doubt. How they will change it is a mystery. In 1995 Robert Allen, at the helm of AT&T, summed up the mixture of bafflement and excitement inspired by this new world:

> One could reasonably expect the chairman of AT&T to know what his corporation will be in ten years from now. He doesn't. One could, within reason, expect the chairman of AT&T to be able to predict how technology will transform his business a decade hence. He can't. At the least, he should know who his major competitors will be in 2005. Stumped again. But here is what he does know: something startling, intriguing, and profound is afoot.[2]

This book attempts to guess what that "startling, intriguing, and profound" something may be. The next three chapters look at the development of the three main technologies at the heart of the revolution—the telephone, the television, and the networked computer. (See Figure 1-1.) Chapter 5 looks at the most immediate impact of change: on commerce and companies. Chapters 6 and 7 look at the policy issues and problems that will face this altered world, addressing such questions as how to ensure competition, whether giants such as Microsoft should be allowed to dominate standards or networks, and what should be done to counter the tendency of new communications technologies to make evil as well as useful knowledge more accessible. The globalization of communications will make it harder to enforce all sorts of national laws designed to protect children, to preserve privacy, or to prevent terrorism. Part of the price of freedom will be a greater need for individuals to take responsibility for their own lives and for the smooth running of society.

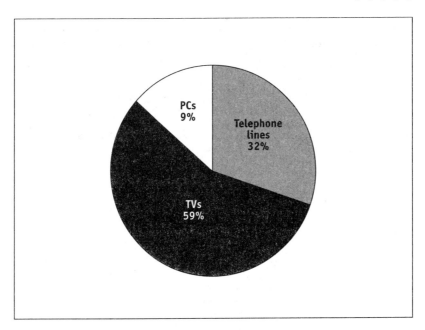

Figure 1-1   Ways to Communicate
Based on a 1995 world total of 2,186 million information devices.
Source: International Telecommunication Union "World Telecommunication Development Report, 1996/97."

The most important questions, covered in the book's final three chapters, concern how the world will change as a result of the communications revolution. The death of distance loosens the grip of geography. Companies will have more freedom to locate a service where it can best be produced, rather than near its market. People—some at least—will gain more freedom to live far from their employers. Some kinds of work will be organized in three shifts based on the world's three main time zones: the Americas, East Asia/Australia, and Europe. Time zones and language groups rather than mileage will come to define distance.

Barriers and borders will break down. The horizontal bonds among people doing the same job or speaking the same language in different parts of the world will grow stronger. Some of society's vertical bonds—between government and the governed, or bosses and workers—will grow weaker. Does that mean that electronic institutions will supplant the institutions that have previously held communities together? That we will have "high-school-in-a-box," say, rather than a real, physical place called high school? Will the personalization of communications go hand-in-hand with social fragmentation? Undoubtedly, governments will find that national legislation is no longer adequate to regulate a global flow of information, even if some of that information is criminal or subversive. Companies will become looser structures, held together mainly by their cultures and their communications networks. For individuals, the lines between work and leisure will grow less distinct. The design of the office and of the home will alter to accommodate the changing patterns of this communications-driven life.

For many people, this prospective new world is frightening. Change is always unsettling, and we are now seeing the fastest technological change the world has ever known. But at the heart of the communications revolution lies something that will, in the main, benefit humanity: global diffusion of knowledge. Information once available only to the few will be available to the many, instantly and (in terms of distribution costs) inexpensively.

As a result, new ideas will spread faster, leaping borders. Poor countries will have immediate access to information that was once restricted to the industrial world and traveled only slowly, if at all, beyond it. Entire electorates will learn things that once only a few bureaucrats knew. Small companies will offer services that previously only giants could provide. In all these ways, the communications revolution is profoundly democratic and liberating, leveling the imbalance between large and small, rich and poor. The death of distance, overall, should be welcomed and enjoyed.

The Roots of Revolution

. . . . .

It is easy to forget how recently the communications revolution began. All three of today's fast-changing communications technologies have existed for more than half a century: the telephone was invented in 1876; the first television transmission was in 1926; and the electronic computer was invented in the mid-1940s.[3] For much of that time, change has been slow, but, in each case, a revolution has taken place since the late 1980s. In order to approach the future, we need first to ask why the really big changes have been so recent and so far-reaching.

The telephone

Since the 1980s, the oldest of the three technologies has undergone two big transformations—an astonishing increase in the carrying capacity of much of the long-distance network and the development of mobility. They result, in the first case, from the use of glass fibers to carry digital signals, and, in the second, from the steep fall in the cost of computing power.

For much of its existence, the telephone network has had the least capacity for its most useful service: long-distance communication. A cross-Atlantic telephone service existed early on: indeed, by the 1930s, J. Paul Getty could run his California oil empire by telephone from European hotels, in which he chose to live because their switchboard operators could make the connections he needed.[4] But even in 1956, when the first transatlantic telephone cable went on-line, it had capacity for only eighty-nine simultaneous conversations between all of Europe and all of North America.[5] Walter Wriston, former chairman of Citibank, recalls the way it felt to be an international banker in the 1950s and 1960s: "It could take a day or more to get a circuit. Once a connection was made, people in the branch would stay on the phone reading books and newspapers all day just to keep the line open until it was needed."[6]

Since the late 1980s, capacity on the main long-distance routes has grown so fast that, by the start of 1996, there was an immense and increasing glut, with only 30 to 35 percent of capacity in use.[7] The main reason for this breathtaking transformation was the development of fiber-optic cables, made of glass so pure that a sheet seventy miles thick would be as clear as a windowpane. The first transatlantic fiber-optic cable, with capacity to carry nearly forty thousand conversations, went on-line only in 1988. The cables that will be laid at the turn of the century will carry more than three million conversations on a few strands of fiber, each the width of a human hair.

Meanwhile, new cables are being laid; new satellites, which carry telephone traffic on less popular routes, are due to be launched; and a range of low-orbiting satellites may eventually carry international traffic between mobile telephones. In addition, new techniques are starting to allow many more calls to travel on the same fiber. It is as though an already rapidly expanding fleet of trucks could suddenly pack several times as many products into the same amount of space as before.

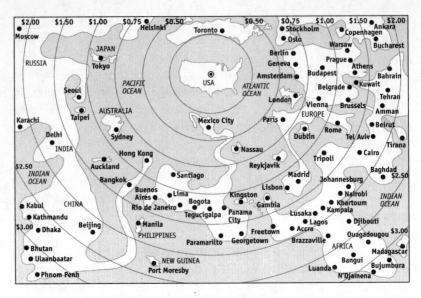

Figure 1-2   The Geography of International Telephone Charges
The cost of making a telephone call from the United States to the rest of the world.
Countries are arranged according to cost per minute of a three-minute, daytime, direct-
dialed international call. The scale of the concentric circles is stated in U.S. cents per
minute. A city that costs $1.40 per minute to call would appear between the $1.00 and
$1.50 circles. Call charges based on June 1994 MCI tariffs and June 1994 exchange rates.
Source: TeleGeography, Inc.—Washington, DC.

This massive growth in capacity is increasingly reflected in tariffs.
MCI's generous Mother's Day gesture cost the firm plenty, but would
have been impossible without the growth in capacity on the American
network, where the traffic on that day is probably the heaviest of any
day, anywhere in the world. Already, international and long-distance
call rates have been falling, changing our mental map of the world, in
ways that are vividly illustrated in Figure 1-2. But the cost of carrying an
extra telephone call across the Atlantic and on many other long-distance
routes has fallen much further and now approaches zero. This fall in
rates is the drive behind the death of distance.

By the middle of 1997, the threat of a glut had receded. The reason
was the enormous increase in demand created by the Internet, which
carries messages of many sorts at prices that ignore distance. When distance carries no price penalty, people communicate more, and in new
ways. In the future, the lavish plans to build more capacity and inge-

nious technologies to compress signals will continue to push prices down, until it costs no more to telephone from New York to London than to the house next door.

While capacity has been increasing, the telephone has become mobile. Cellular communication, which dates back to the period immediately following World War II, became commercially viable only in the early 1980s, when the collapse in the cost of computing made it possible to provide the necessary processing power at a low enough cost.

Now, the mobile telephone may arguably be the most successful new way of communicating that the world has ever seen—already, more than one telephone subscription in seven is to a mobile service. Mobile telephony's share will continue to rise: in 1996, it accounted for 47 percent of all new telephone subscriptions.[8] For conversations, people will come to use mobile telephones almost exclusively.

They will be able to communicate from every corner of the globe: in the course of 1996, two stranded climbers on Mount Everest used mobile telephones to call their wives. One wife, two thousand miles away in Hong Kong, was able to arrange her husband's rescue; the other, sadly, could merely say a last farewell.[9]

The mobile telephone also allows better use of the most underused chunk of time in many peoples' lives: traveling time. People will use their commuting time more fully, but other benefits may be even greater: passengers can be checked in for flights during the bus ride to the airport, for example, and maintenance staff can schedule visits more efficiently, knowing exactly when equipment in transit will arrive. The mobile telephone thus raises productivity by using previously idle time.

The television

At the end of the Second World War, a mere eight thousand homes worldwide had a television set. By 1996, that number had risen to more than 840 million—two-thirds of the world's households.[10] The basic technology of television sets has not changed over those fifty years, but the transmission of programs has been revolutionized by the development of communications satellites. Now another revolution—in channel capacity—has begun.

In fall 1963, people around the world witnessed for the first time an important but distant political event as it was taking place. The 1962 launch of Telstar, the first private communications satellite, had made possible the live global transmission of the funeral of President John F. Kennedy.[11] The psychological impact was huge: this unprecedented new link among countries would change perceptions of the world, creating the sense that the world's peoples belonged to a global, not merely local or national, community.

The 1988 launch by PanAmSat of the first privately owned commercial international (as opposed to domestic) satellite, constituted another milestone, cutting the cost of transmitting live television material around the world. As recently as the 1970s, more than half of all television news was at least a day old. Today, almost all news is broadcast on the day it occurs.[12] Big events—the fall of the Berlin Wall, the Gulf War, the O. J. Simpson trial verdict—go out to billions of viewers as they happen.

Until recently, most television viewers around the world have had access to perhaps half a dozen television channels at most—and often to only two or three. The main reason is purely physical: analogue television signals are greedy users of spectrum. Only in the United States and a handful of other countries, and mainly only since the 1980s, have cable-television networks—less constrained by the limits of spectrum—brought people real viewing choice.

Now choice is expanding with breathtaking speed. Toward the end of the 1980s, communications satellites began to broadcast directly to a small dish attached to people's homes, thus inexpensively distributing multichannel television. Suddenly, more viewers had more choice than ever before.

In the mid-1990s came another revolutionary change: broadcasters began to transmit television in digital, not analogue, form, allowing the signal to be compressed and, consequently, far more channels to be transmitted, whether from satellite, through cable, or even over the air. Like the long-distance parts of the telephone network, a service that had been constrained by capacity shortage for most of its existence has suddenly begun to build more capacity than it knows what to do with.

The result will be a revolution in the nature of television. For those who want it (most of us), the old passive medium will remain, a relaxing way to pass the evening after a day spent at work. But television—the business of transmitting moving pictures—will develop many more functions, including new roles in business. The finances of television will also change, and in a way that many viewers will resent. The scarcest thing in television is not transmission capacity, but desirable programs, especially live programming. In the future, these will rarely be available at no cost to viewers. Increasingly, viewers will pay directly for what they most want to watch.

The networked computer

The newest of the three building blocks of the communications revolution, the electronic computer, has evolved fastest. In 1943 Thomas Watson, founder of IBM, thought that the world market had room for about five computers.[13] As recently as 1967, a state-of-the-art IBM, costing $167,500, could hold a mere thirteen pages of text.[14]

Two key changes have altered this picture. First, computing power has grown dramatically. As a result, the computer can be miniaturized and has become a consumer durable, with computing power embedded in everything from automobiles to children's toys. The main processor on Apollo 13 contained less computing power than does a modern Nintendo games machine.[15] Second, computers are increasingly connected to each other. The Internet, essentially a means of connecting the world's computers, makes apparent the spectacular power of such networked computers.

The increase in computing power has followed a principle known as "Moore's Law," after Gordon Moore, co-founder of Intel, now the world's leading maker of computer chips, the brains of the modern computer. In 1965, Moore forecast that computing power would double every eighteen months to two years. So it has done for three decades, as engineers have found ways to squeeze ever more integrated circuits of transistors onto chips—small wafers of silicon. A 486 chip, standard in a computer bought around 1994, could perform up to fifty-four million numerical calculations per second. A Pentium chip, the standard three years later, could perform up to two hundred million calculations per second. And Moore's law continues to apply. By 2006, according to Intel forecasts, chips will be one thousand times as powerful and will cost one-tenth as much as they did in 1996.[16] (See Figure 1-3.)

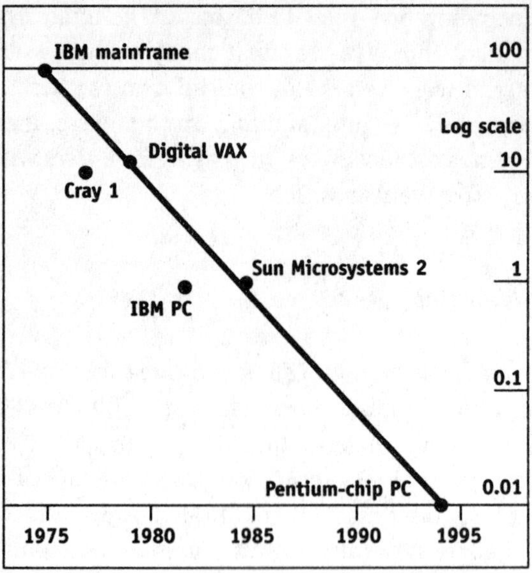

Figure 1-3   Moore's Law in Action
Cost of information processing in dollars per instruction per second, 1975 = 100.
Source: *Global Economic Prospects and the Developing Countries 1995*, World Bank.

As the power of the chip has multiplied, the price of computing power has fallen, computer size has decreased, and computer capacity has risen. This has had implications for many aspects of communications: the development of mobile telephones, for example, and of the "set-top boxes" that decode encrypted television signals. A landmark occurred in 1977, when Steven Jobs and Stephen Wozniak, two young computer enthusiasts, launched the Apple II, opening the way for the computer to become a household good. Today, 40 percent of homes in the United States contain a computer.

Meanwhile, responding to the limitations of computers in the 1960s, when they were large, expensive and scarce, the American Defense Department's Advanced Research Projects Agency (ARPA) backed an experiment to connect computers across the country as a way to exchange messages and share their processing power. This effort yielded a nationwide network that initially linked only university computers. Because different computers in those early days had different operating standards, a common standard or protocol became a fundamental requirement of the network. In response, Transmission Control Protocol/Internet Protocol (TCP/IP) was developed and, since the early 1980s, has provided the format for packaging all data sent over the Internet.

TCP/IP is the essence of the Internet. It provides an electronic Esperanto: a common language and a set of rules through which computers all over the world can talk to one another, regardless of whether they are PCs or Apple Macs, whether they are vast university mainframes or domestic laptops. Through the Internet, any number of computer networks can connect with one another and behave as a single network.

Although the use of the Internet grew rapidly in the 1980s and early 1990s, doubling every year, its transformation into a popular success dates only from about 1993 to 1994. At that point, the World Wide Web made it possible to accommodate on-line graphics, sound, and moving pictures, rather than just text, making the Internet more versatile and more interesting to look at. This was thanks to Marc Andreessen, a young programmer, and his colleagues at the University of Illinois. They developed the most successful graphical Web browser, which allowed navigation fairly easily from one screenful (or "page") of information to another, even if that second page was held on a different computer in another part of the world, simply by using a hand-held "mouse" control to point and click on a shaded word on the screen.

These transformations have had three main consequences. First, they vastly increase the world's computing power. Even if Moore's law stopped grinding inexorably along, the Internet has the potential to allow immense multiplication of computing power simply by linking many different computers. Second, the Internet has emerged, almost by accident, as the first working model of the "global information superhighway" that politicians and big communications companies talked so much about in the early 1990s.[17] It has become not only a new global means of communicating but also a new global source of information on a gigantic scale. Third, the Internet has given birth to a vigorous new industry dedicated to developing ways to use it and services to sell across it. Only in 1994 did the number of commercial computers connected to the Internet overtake the number of academic computers. Now, tens of thousands of companies, many of them small start-ups, are racing to find profitable uses for this new technology. Never in history have so many entrepreneurs attempted, in so short a space of time, to develop uses for an innovation.

The Internet is thus a global laboratory, allowing individuals as well as the marketing departments of multinationals and academics in top universities to pioneer uses for communications technology. Already it carries telephone and video conferences as well as live television and radio broadcasts. All sorts of communications experiments, carried out on the Internet, will feed through into the other media, changing and developing them. The Internet thus functions as both a prototype and a testing ground for the future of communications. Watching its evolution, we can catch a glimpse of what lies ahead.

Competition and Its Consequences
· · · · ·

For people to reap quickly the benefits brought by the communications revolution, competition will be essential, because competition is the best guarantor of choice, quality service, and low prices. Take the example of the car industry. Countries with little competition, such as Russia and India, tend to have few car producers and few imports. Not only do they fail to produce good cars, but the liberating impact of the automobile has largely passed them by.

That may sound obvious. But the telephone and television industries have been amazingly protected by comparison with, say, automobiles or other consumer durables. Around the world, most people have had to buy their local telephone connection from the only company available; only one, or at most a handful, of companies have had the right to carry calls across international borders; and choice in television has been limited. Outside the United States, most countries have had a state-owned national telephone giant and often a state-owned television giant, as well.

Two factors are changing this communications picture. First, technology, which once imposed constraints on competition in telephone and television networks, is now making competition easier by massively increasing the number of potential delivery channels. Second, some governments are keen to foster competition, because they see that consumers will benefit from it, and that the country will suffer without it.

So competition will eventually happen. But deregulation raises two important policy issues. The first concerns the way these services are delivered. How can countries achieve genuine competition? In particular, how can they ensure that a single private monopoly does not come to dominate the industry in the future as has happened in the past? Communications businesses are prone to concentration: they create giants such as IBM and Microsoft as well as monsters like Japan's NTT and Germany's Deutsche Telekom. To cope, governments have to go from being restrictive regulators to being active promoters of competition.

The second policy issue concerns the services that actually flow over the new networks. In the past, it has been easy to ensure that, for example, child pornography is barred from television screens, because television has been provided by a monopoly or a handful of companies. But in the future, patrolling the frontier will be much harder. Repressive governments will find it much more difficult to keep free ideas off the airwaves, but democratic governments will also find it far harder to prevent on-screen child pornography, racism, fraud, or libel.

The influence of technology

Two changes will particularly help to ensure more choice. First, many of the costs of providing a service, whether a telephone calling-card service, a new television program, or a news service delivered over the World Wide Web, have fallen steeply. That has allowed many new companies to invent services for customers. Second, the same service can be delivered—to the home or the office—through more than one channel. Already, a telephone call can be received on an ordinary handset or over the Internet using a personal computer. Using specially designed sets, people can both watch television and explore the Internet.

Both these changes spring from a crucial fact: the telephone, the television, and the networked computer constitute three different ways to handle material in digital form. Computers have always handled information this way: they work in the single stream of ones and zeroes, or pulses and lack-of-pulses, that is digital code. As the capacity of computing has grown, it has become possible to convert into digits information that would once have been transmitted in analogue waves, such as music, speech, and moving pictures. First the telephone network and now, increasingly, television networks are switching from analogue to digital formats.

The resulting electronic common currency blurs the traditional divide between media. Once information is handled digitally—whether a Hollywood blockbuster or a telephone conversation—no technical requirements mandate a special machine for each task (although there may still be an ergonomic need). Instead, digital information can be sent from one computer to another—or from a computer across a telephone network to the set-top box of a television. A personal computer connected by a modem to the telephone network can already transmit video pictures (although they may be small and jerky) and telephone calls (although the sound quality at times would make even Alexander Graham Bell wince).

Thus, competition of all sorts is becoming easier. Around the world, telephone companies are experimenting with sending television signals over their wires, and cable-television companies are carrying telephone calls on their networks. So far, it must be said, companies have found it difficult to transfer their skills: for instance, the telephone companies have had a bad time in the television business. More important, in the long run, will be the way that both industries can be challenged by upstarts on the Internet.

That challenge will grow as it becomes possible to overcome the biggest technical barrier to competition—the absence of a high-capacity two-way link to homes and small businesses. At present, twisted pairs of copper wire, a technology designed for carrying the human voice and unchanged for 120 years, constitute the usual local connection. Some homes also have cable-television links, designed to work in one direction only. Both technologies are unsuited to carry Internet traffic, which requires plenty of capacity (so that those video pictures look less jerky) and a two-way flow.

A huge effort is now being made to find ways to increase capacity inexpensively: by using digital compression to squeeze more information through ordinary telephone lines, for instance; by using wireless to a fixed receiver, instead of a wire connection; by replacing copper with fiber-optic cable; by adapting cable-television networks; and by using combinations of satellite and telephone. Over the first decade of the next century, measures such as these will allow homes and small businesses much faster access to the Internet.

The influence of government

In most countries, governments have run the telephone service; in many, they own or finance at least one television channel. This pattern has encouraged them to suppress competition. Now, many have put their telephone companies up for sale, and more and more are allowing—or even encouraging—competition. That is good news. In the long run, governments may not be able to stop competition from happening, but they can certainly hold it back for a decade or two if they really try.

In contrast, in the United States, the Telecommunications Act of 1996 allowed long-distance and local telephone companies and cable-television operators to enter one another's markets (although court cases have been a hindrance). In Britain, cable-television companies have been providing telephone services since 1991. Under a deal negotiated at the World Trade Organization in early 1997, several countries will start to open their telecommunications markets to foreign competitors early in 1998. The richer countries of the European Union had already agreed on that deadline for allowing competition. Quite a few developing countries are starting to see that they win more than they lose by opening their markets and even by allowing foreign companies to build and run their telephone networks.

In fact, competition will need government help to become established. Big networks are more useful to customers than small ones— think how irritating it would be if your mobile telephone could reach only other people on the same service, rather than any telephone subscriber in the world. So newcomers—such as, perhaps, your mobile-telephone service provider—have to bargain with the big telephone monopolies, companies such as Bell Atlantic or BT or Deutsche Telekom, for permission to link into their networks. The cards are all stacked against them—unless government intervenes to set fair rules.

The protection problem

Even the most liberal societies have had rules about the way information is made publicly available. Some of the rules are to prevent people saying or showing illegal things—such as child pornography or fraudulent advertising or racist abuse. Others are to protect the ownership of ideas and other creative endeavors from intellectual piracy. Experience with the Internet suggests that all of these rules will be harder to apply, for two reasons. First, the Internet allows much more information to flow freely across borders. With the Internet, a computer user in the United States can look at material held on a computer in, say, the Netherlands that would be prohibited if it appeared in an American television program or magazine. The Internet's ability to cross borders makes it difficult to regulate using only national laws. All sorts of material can travel over the Internet in ways that escape national laws on gambling, pornography, racism, fraud, libel, terrorism, and taxation. Second, the Internet blurs the distinction between the public and the private spheres. Television broadcasts have been regulated because they are a public medium; what people say in a telephone call is almost as unregulated as private conversation. The Internet blurs that distinction too: the same medium can deliver a private communication (such as electronic mail) and a public one (such as a Web broadcast or a visit to a Web "site") over the same connection and to the same screen.

In the past, the three communications industries have all played by different rules, often policed by different regulatory bodies. The software business has been regulated like most other industries, which is to say hardly at all; television companies have had complicated rules about who can own them and what they can or cannot show on the screen; while telephone companies have had rules about the way they offer their service and the prices they can charge.

For the moment, the three industries, of necessity, will maintain their separate identities: television is not interactive, the Internet is a mediocre telephone, and the telephone is not designed to receive video pictures. But that will change. When it does, the three industries will have to be regulated in much more similar ways—if at all.

The Long-Term Effects

· · · · ·

Imagine the automobile in 1910. Twenty-five years after its invention, it had already assumed a form recognizable today: it had a gasoline engine at the front, pneumatic tires, a clutch and a gear box, and its most advanced models could travel at today's highway speeds. Most important of all, Henry Ford had started to mass-produce it, making it clear that the automobile would become a normal consumer item.

Yet none of the social consequences of its development were apparent. The physical landscape had yet to adjust: highways and out-of-town shopping centers did not exist; suburbs sprawled only around railway stations; and city centers were still lively hubs of commerce and industry. Not only have the geographic consequences of the automobile taken most of the century to unfold, so have the economic and social consequences of increased personal mobility. The automobile destroyed many jobs (who needs ostlers now?) even as it created millions more. It built new industries and created markets for new skills (truck driving, repairs). It made possible the flight of the middle classes, particularly in North America, from the inner city. And it liberated ordinary people (including women) to travel where and when they wished.

The communications revolution will be no different. The telephone, the television, and the computer have already transformed the economy and society in ways unimagined half a century ago. The telephone has given birth to new jobs: a sign of wealth is no longer having a butler to answer your door but a secretary at home to filter your calls. The telephone has built a vast industry: in most countries, the telephone company is one of the biggest employers. And it has brought commercial change, allowing companies to operate globally, rather than merely nationally. Financial markets, the quintessentially global industry, have grown up on global electronic communications.

The changes brought by television have been every bit as great. Activities with which people once filled their time—reading newspapers or writing letters, for example—have dwindled in importance. The power of television has transformed not only social life but politics and entertainment. For the first time in history, most people know what their politicians look like (and choose them partly based on their television manner and image). Like the automobile, television has created demands for new skills and has been the basis for new industries. It is, with radio, the foundation of the modern advertising industry—and of the concept of the brand, whose emergence, in the 1950s, paralleled the main growth in television ownership in the United States.

The changes brought by the computer are less evident, at least in economic terms: indeed, one of the mysteries of the final quarter of this century is why so much spending by companies on computers has pro-

. . . . .

duced so little return in higher economic productivity. In homes, too, the huge sales of personal computers—rapidly drawing level with sales of television sets in some countries—has not yet changed social behavior as strikingly as the telephone and television have done. Networked computers will be another matter. They transform the vast number of interactive activities—sales calls, management meetings, information gathering, problem solving, reporting and communicating—that are the bedrock of economic activity. The transformation in the speed and capability of such interaction will revolutionize the productivity of up to half the work force of the rich world.[18]

Key questions for the future

Now a second generation of change is taking place, bringing with it a broader range of questions. How will demand for communications technologies evolve? Will the Internet continue to develop as a communications medium in its own right, or will it be used to enhance the television and the telephone? Will people watch television on their office computer screens or use the television at home as a gateway to the Internet?

Beyond that, how will these technological changes alter the world? The communications revolution will transform how we work and shop and how we are governed; our health care and education, our leisure and social lives; and the shape of companies and of cities and the design of our homes. What really matters are the effects on the economy and society, for inventions can intimately affect the pattern of people's lives.

The evolution of demand

Guessing the way demand for a technology will evolve is a game with a high failure rate. Even those close to an innovation may fail to spot its significance. A McKinsey study of AT&T in the early 1980s averred that "the total market for mobile cellular phones will be 900,000 subscribers by the year 2000."[19] In fact, by 1996 there were already one hundred times as many. On that advice, AT&T pulled out of the market, only to re-enter it at great expense through the purchase of McCaw in the mid-1990s. Almost everybody sometimes gets technology wrong.

With that proviso, here is a guess. The next few years are likely to see the increasing integration of the Internet with both the telephone and the television. The most immediate effect of the communications revolution will be to allow familiar objects to be used in new and unfamiliar ways. The telephone and the television will both incorporate mobility and Internet technology in ways that will extend their use and increase their versatility. The most powerful corporate role will be played by the telephone companies. Largely free from debt, with powerful national brands, the telephone companies have the muscle and ubiquity to exert enormous influence over the future of communications, at least in the short run.

They may well become the main providers of Internet services, offering users a premium service—for a fee—that will be fast, reliable, and secure. They will rarely be innovators, but they will drive forward the commercial use of the Internet, building links between the Internet and the telephone so that customers can move swiftly between screen and voice contact or use both together in the same transactions. The incorporation of Internet technology will thus transform the way the telephone can be used. In addition, the global toll-free number will allow the combination of the telephone and an Internet site to be used as a storefront anywhere in the world.

Integration of the Internet and television will occur differently. The main impact on television will be the way people use their leisure time. As an activity that absorbs time, rather than saves it, television will be vulnerable to competition from other inexpensive communications, such as a long telephone call to a friend or a session playing on-line games.

In response, television will become the source of several different services. It will still offer popular shows and entertainment, although the concept of a "channel" with a set timetable for shows may disappear: people will choose when they will watch a show. One result will be to increase the overall audience for the most popular shows—and their revenue, too, since people will increasingly pay to watch the most desirable content. But a second sort of television will embrace the Internet to offer services such as weather forecasts and traffic reports individualized and tailored to a viewer's location and interests. Channels will become more interactive, merging with Internet content, such as chat groups or electronic games. They may become visual versions of radio talk shows: participants will be able to see one another, even show one another their home videos, on screen. Some such programming will be moderated and have contributions from experts; others will be a jumble of amateur participation.

Apart from enhancing the telephone and television, the Internet will grow mainly through "intranets," private networks that use its common language, established by companies for internal communications, and "extranets," to communicate with suppliers, distributors, and corporate customers. The premium services offered by the big telephone companies will almost certainly have some new name, for the companies that provide them will not want to be too closely associated with the chaos of the public Internet. But that will remain and continue to flourish, a hubbub of creativity, communication, congestion, and crime. Freely accessible to anybody who wants to put anything on it, the Internet will remain a breeding ground for any number of radically new services.

Patterns of adoption

It takes time for a new technology to find its true markets. Demand for the personal computer at home boomed once people began to use it as a superior games machine. Just as people initially thought of the automobile as a horseless carriage or saw television as radio with pictures, so today's Internet users are mainly using this new medium for the same old services.

It takes time, too, for new business models to emerge. People are used to paying directly for telephone service (although, with toll-free calls, a growing share of the cost is carried by companies), and they are used to the idea that advertisers pay for broadcast television. But the payment structure for the Internet is still developing. It is still not clear whether users or advertisers will pay, or, as with cable television, whether payment will be made by some combination of the two.

In the short and medium term, the way people adopt new technologies will be influenced by culture, convenience, and cost. It will also vary enormously from one country to the next.

The following factors will be among those that determine how each technology is used in the future:

- Culture. If people are used to doing things one way, they may be slow to change—especially if, like one-fifth of the rich world's population, they are over 65. Some of the developing countries, with their younger populations and fewer preconceptions, may thus be quicker than the older West to spot and take advantage of new possibilities.

- Convenience. Some technologies can be readily used in one situation but not in another. It is for instance, possible to listen but not to watch while driving an automobile; it is easier to switch on the television than to find a Web site. The more effort it takes to master the use of a technology, the slower and more limited its diffusion will be. Hence the personal computer will never be as widely used as is the television.

- Cost. Demand will be boosted when new technologies cut costs. The Internet has flourished most in the countries where access is least expensive. In addition, the Internet itself will lower costs— for instance, of advertising and promoting many services. The services whose costs will be most affected will be those that can be electronically distributed, from computer software and videos to financial services and information.

The strength of these factors will vary from one country to another. The penetration of communications technologies varies widely, even within the rich world. The fixed telephone and the television are virtually universal. But there are still big differences. In Sweden, where cellular telephone services began in 1981, more than one person in four has a mobile telephone — making mobile phones more than one and a half times as common, relative to population, as in the United States. In France, where GNP per head is much the same as in Sweden, the figure is one in twenty-four. In Canada, three-quarters of homes have multi-channel television, delivered by cable or satellite. In Australia, perhaps the country with a culture most similar to Canada's, the figure is about one in twenty. In the United States, there are thirty-eight Internet "host" computers per thousand people. In Japan, a country whose consumers usually adopt new technologies almost as rapidly as do Americans, there are only six.

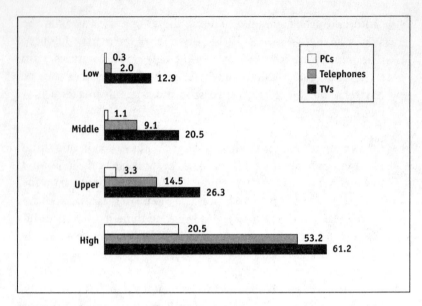

Figure 1-4   How Access Varies with Wealth
Number of television sets, telephone lines, and PCs per 100 inhabitants, 1995.
Source: International Telecommunication Union "World Telecommunication
Development Report, 1996/97."
Notes: Income groups: U.S. $ GNP per head: low = $725 or less, middle =
$726–$2,895, upper = $2,896–$8,955, high = $8,956 or more.

The differences between developed and developing countries are
greater still. (See Figure 1-4.) Almost one-third of the world's population lives in countries that, added together, have fewer than nineteen
million telephones among them—fewer than in Italy. The whole of
Africa, with its population of five hundred million people, has as many
telephones as Tokyo. But these numbers are changing swiftly, as new
ways of providing telephones, such as wireless telephony, cut the cost
and allow lower-income countries to leapfrog several technological
stages. By around 2020, if not sooner, many developing countries will
have caught up. Their communications will be as good as those of many
of what are now the world's richer nations.

Out in front of the pack, though, will be the United States. What is happening to communications plays particularly to American strengths. The
United States has the right history. It has long had competition in parts of
the telecommunications market; many countries are only now starting to

allow competition. America has long had multichannel television; for most countries, that is relatively new. The Internet itself was developed in the United States. In addition, low prices for leased telephone lines (which carry Internet traffic), low telephone tariffs (with local charges not usually based on the length of a call), and wide ownership of personal computers have all helped to drive forward the use of the Internet.

Add to this the following cultural factors: the intellectual-property industries, such as movie-making, software, and popular music, have traditionally been important. The primary language of the Internet is English, now the world's most widely spoken second language. The geographic size of the United States has made people and businesses comfortable with long-distance commerce, such as mail-order shopping and credit-card ordering. Americans typically show interest in new ideas and a general tendency to adopt innovations early. And Americans have traditionally been willing to take commercial risks, making venture capital accessible.

All these factors will give the United States a lead in communications and its uses that other countries will find impossible to beat. As a result, the first half of the next century, even more than the second half of the present one, will be dominated—economically, politically, and culturally—by the United States.

Changing the World

· · · · ·

"The full importance of an epoch-making idea is often not perceived in the generation in which it is made," observed Alfred Marshall, one of the fathers of modern economics, in his *Principles of Economics*. "A new discovery is seldom fully effective for practical purposes till many minor improvements and subsidiary discoveries have gathered themselves around it."[20] Not until the middle of the next century will it be quite clear what the broadest impacts of new communications have been. The effects will be felt in four main areas: commerce and the shape of the company, the economy, society and culture, and government and the political process.

Commerce, including many kinds of retailing, will become increasingly international. Armed with a credit card, the nearest thing we have to a world currency, people will eventually shop around the globe. The big barriers, such as customs formalities, delivery costs, and differing rules of consumer protection, will gradually decline. Such global retailing will allow many niche markets to emerge.

Companies will be able to build new links not just with customers but with employees in different parts of the country—or around the world. Employees in different countries or regions will be able to work in teams on the same project. But a bigger consequence will be the reduction in the size of firms in many industries. Computers and communications will allow companies to become networks of independent workers, specializing in what they do best and buying in everything else. Employees will, therefore, often work in smaller units or on their own.

All kinds of services will be bought in. As communications improve, it will become easier for companies to hunt for spare capacity and low prices. A knowledge-based company can buy in much more of what it needs—design, say, or marketing or packaging—than can a traditional company. There will be new opportunities to integrate customers and suppliers, using the corporate communications network as the connective tissue. Suppliers will tap into their customer's electronic data base to work out what the customer needs and to supply it—just as smoothly as an in-house supplier would once have done. Some suppliers will serve many customers, some only one. Technologies such as electronic mail and computerized billing will reduce the costs of dealing with suppliers at arm's length. The result of intensified competition will be a greater emphasis on performance, productivity, and waste reduction.

Economies will benefit from the vast increase in the diffusion of knowledge and information, the basic building blocks of economic growth. This revolution will be especially important because innovation—the introduction of new production methods, new products, and new kinds of industrial organization—is the main force driving growth and thus living standards. The communications revolution speeds up the diffusion of innovation. It will thus allow new competitors to spring up, companies to react quickly, and individuals to spot opportunities. Information is also essential to make markets work well. Many more companies will have access to accurate price information, wiping out excessive profits, enhancing competition, and helping to curb inflation.

For many products, global markets will become more important than local or national ones. The change will take time: international trade in services, far more than trade in goods, is shackled by all sorts of regulatory barriers. But, eventually, trade and foreign investment will become even more important, and economies even more interdependent, than they are now.

More companies will become footloose—more willing to locate wherever the best bargain of skills and productivity can be had. The result will be to calibrate wages with productivity more precisely, in more industries, by world standards. That in turn will raise the global premium for skills. The result may be greater pay inequality within countries, but more similarity among the pay of different groups similarly employed around the world.

Society, too, will change, with the disappearance of the old demarcation between work and home. More people will work from their homes or from purpose-built small offices. The office may become the place for the social aspects of work, such as networking, lunch, and catching up on gossip, thus inverting the familiar roles of home and office. City centers will lose some of their present function as office centers and instead develop as centers of entertainment and culture: places where people go to stay in a hotel, visit a museum or gallery, or go out for a meal or to hear a band. The result may not necessarily be less traveling—indeed, when people need to meet their electronic contacts, they may actually travel farther than before—but the traveling will be of a different kind.

Many people, especially in continental Europe, fear that electronic communications will erode their cultures and replace them with something Anglo-Saxon and Hollywood-driven. In fact, the decline in the cost both of creating and of distributing many entertainment products is more likely to reinforce such cultures than to undermine them. French cinema will never again rival Hollywood, but the multiplication of global television channels will create many more outlets for French films.

Politics and government will be transformed by free communications, changing the balance of power between governments and their citizens. People will become better informed and will be able to communicate their views to their government's leaders and representatives more easily. Politicians will become more sensitive to lobbying and to public-opinion polls, especially in the established democracies. People who live under dictatorial regimes will find it easier to communicate with the rest of the world.

The size of the nation state will also be affected. The death of distance will not only erode national borders; it will reduce the handicaps that have up until now burdened fringe countries. That will be of enormous importance for the many small countries that have come into existence in the past half century. As a result, one economic argument against secession will be eroded.

Above all, the death of distance will be a force for global peace. Where countries and their citizens communicate freely, they will surely be less likely to fight one another. Individuals everywhere will know more about the ideas and aspirations of human beings in every other part of the world, thus strengthening the ties that bind us all together.

"And in today, already walks tomorrow," wrote Samuel Taylor Coleridge. Many people worry about this electronic future. They see the prospect of many jobs destroyed and a few sets of skills disproportionately rewarded. They worry about society's increasing vulnerability to technological breakdown and computer crime. The more society relies on technology, many argue, the bigger the problem when something goes wrong and computer systems are hacked into or go haywire.

But many of these worries already apply to the non-electronic world. The water supply, the freeways of Los Angeles, the Tokyo subway—all make society more vulnerable than before to sudden disruption, but in exchange for benefits so vast that we accept the risks. And against the danger of jobs destroyed, we must set new opportunities—including opportunities for some who lack them in the non-electronic world. "On the Internet," says a famous *New Yorker* cartoon, "nobody knows you're a dog"—but nobody knows, either, whether you are young or old, black or disabled, a man or a woman.

This is a revolution about opportunity and about increasing human contact. It will be easier than ever before for people with initiative and ideas to turn them into business ventures. It will be easier to discover information, to learn new things, to acquire new skills. Above all, it will be easier to find somebody to talk to—to communicate, whether with friends or strangers, relatives or customers. As a result, the world will, in all probability, be a better place.

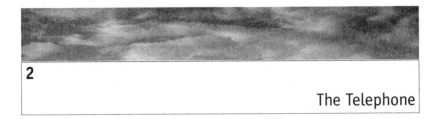

2

The Telephone

The telephone lies at the heart of the communications revolution. In its existence of a century and a bit, the telephone has already transformed social and economic life beyond anything that Alexander Graham Bell might have dreamt of back in 1876. It has brought companionship, employment, and information to millions. Now this familiar instrument has become the gateway to a new world. What has happened? And where will it lead?

The revolution now taking place has three aspects: the collapse in the cost of providing a long-distance call, the increase in the number of things a phone can do, and the wiring of the world. Because the telephone is so versatile and so widely used, these changes will have profound effects. The death of distance will be driven by the plummeting price of long-distance and international calls. The price has fallen a long way since 1927, when the first residential telephone service opened between London and New York at the princely cost per three-minute call of £15. (See Figure 2-1.) But this is just the beginning. Eventually it will cost no more to telephone from Hollywood to London than to telephone to nearby Beverly Hills. The result will be a world where distance, for the first time in history, imposes no barrier to communication.

At the same time, the telephone will acquire new capabilities. Already it is portable. Increasingly, it can provide a soaring number of extra services. The black Bakelite instrument with a dial that sat on a table in the hall and rang just as you were getting into the shower can now be used like a small but immensely powerful computer. Connected to the telephone network, it can contact millions of other computers around the world. In effect, it is becoming one of the main entry

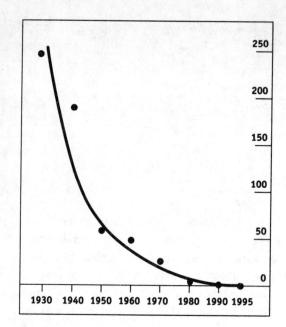

Figure 2-1   The Falling Price of a Three-Minute Call
Price of a call placed from New York to London in 1990 dollars.
Source: *Global Economic Prospects and the Developing Countries 1995*, World Bank.

points to the Internet. Its future will be shaped by the ways it makes use of the Internet.

Meanwhile, the wiring of the world goes on apace. In 1996, 5.8 billion people made do with a mere 745 million telephone lines. Most of those are in the rich world: 80 percent of the world's people share only 30 percent of its telephone lines. Over the next quarter century, that will change, as most developing countries take advantage of new ways to provide telephones at lower cost. The countries of the developed world will thus lose one of their comparative advantages over the less developed countries: vastly better communications. This change could have consequences as significant to world economics as did the collapse of the communist system.

The forces for change are not just technological. In addition, privatization and deregulation are bringing competition into the telephone industry—in many countries, for the first time. The richer countries of the European Union have pledged to open their markets to competition at the start of 1998 and, thanks to a World Trade Organization agreement reached in early 1997, so have a number of other countries. Making competition work will be difficult, but at least this is a start.

Competition will be good for customers: new services emerge more often where competition is lively. Business customers will benefit first and most, but individuals, too, will eventually enjoy most of the concessions that companies receive. Some of the new services will build on the technology of the Internet, as described in the next chapter.

These changes will reshape the telephone business. From being dominated by monopolies—a single state-owned giant or (in the United States) a few big regional monopolies and a long-distance leviathan—telecommunications is becoming more like other fast-changing industries: full of newcomers, including newcomers from other countries and from other industries. It will retain its tendency to concentration, but today's big telephone companies will find themselves facing unprecedented commercial pressure.

The telephone still has enormous undeveloped potential. Despite all the talk about the information superhighway and the advent of multimedia, what happens to the telephone itself will have more influence on people's lives through at least the first decade of the next century. Not only is the telephone roughly three times more common, even in the developed countries, than the personal computer,[1] it can already be used to check a bank account, take part in a radio show, sell insurance, lobby a politician, chat up a porn queen, check messages, or call a friend. Now, the cost, flexibility, and range of this wonderful invention are being transformed and, with them, the lives of those who use it.

The Price of a Call
· · · · ·

In almost every country, ordinary telephone customers pay more—often scandalously more—for a long-distance or international call than it costs the telephone company to carry it. That is changing, but relatively slowly. In the first decade of the next century—sooner in some countries—it will change faster. Eventually, no extra charge for duration or distance will be made for most telephone calls.

The fall in the premium for distance carries a cost. The large profit earned on long-distance and international calls, most of them made by commercial and business telephone users, have in the past cross-subsidized local calls, especially those made by individuals. As that cross-subsidy goes, local call rates will not decline: in fact, they may initially rise. For many individuals, the benefits of the death of distance will therefore be slow in coming. What will come about, however, will be a closer alignment of the true cost of a call and what users pay.

The disappearing distance premium

The telephone companies (and, indeed, their customers) "grew up" with the idea that capacity was most limited on the international and long-distance parts of their networks. Hence the premiums charged for distance. Now, scarcity has given way to plenty. The costs of these parts of the network have fallen by about 25 percent annually for the past two decades. But prices have been slow to follow, especially in countries with little competition.

The main force driving prices down has been the increase in capacity and the fall in the cost of providing it. In the first half of the 1990s, capacity grew much faster than demand. New fiber-optic cables—the oil pipelines of the information economy—are still being laid at a tremendous rate. More than two dozen already snake their way under the world's oceans. They have been laid most lavishly under the Atlantic and Pacific and along busy routes in the United States. In addition, almost all the American long-distance network now uses fiber-optic cable.[2]

The new cables to be laid in the mid-1990s are capable of carrying more traffic than all the previous network of submarine cables combined.[3] At the same time, capacity on the other main long-distance technology, the satellite, is also racing ahead. (See Table 2-1.) Much of this new capacity will eventually be filled as the growth of the Internet and falling prices create booming demand. But the important point is that building capacity now costs much less than it once did, and each cable or satellite has far more capacity than ever before. The capital cost of constructing one transatlantic cable voice path—the capacity to carry one telephone conversation—is now around $1,000. Over the lifetime of the cable, that works out to less than one hundredth of a U.S. cent per minute of use. That figure

Table 2-1   Cable and Satellite Capacity on Transatlantic and Transpacific Routes, 1986–2000

Year	Transatlantic (North America–Europe) Voice Paths		Transpacific (North America–East Asia) Voice Paths	
	Cable	Satellite	Cable	Satellite
1986	22,000	78,000	2,000	39,000
1987	22,000	78,000	37,800	39,000
1988	60,000	78,000	37,800	39,000
1989	145,000	93,000	37,800	39,000
1990	145,000	283,000	37,800	39,000
1991	221,000	283,000	114,200	27,000
1992	296,600	496,000	190,500	27,000
1993	410,000	620,800	264,000	83,300
1994	701,800	620,800	264,000	234,000
1995	1,310,800	710,800	264,000	234,000
1996	1,310,800	710,800	864,600	234,000
1997–2000*	1,310,800	737,500	1,464,600	424,500

*Minimum available
Source: TeleGeography, Inc.—Washington, DC.

will fall even further, as new compression techniques squeeze more onto each fiber. With operating costs added, the cost per minute for a satellite is around 0.003 cents, even if, on average, only a quarter of its capacity is in use.[4] The basic cost of carrying an additional telephone call on the transatlantic route is thus as near to zero as makes no difference.

Of course, there are other costs. The switching that routes a call is increasingly sophisticated. More important, the costs of the local loop—the last part of a telephone network, connecting the customer to the main system—have fallen only slowly. In addition, telephone companies tend to employ lots of people for activities such as billing and marketing. NTT, Japan's telephone giant, and Germany's Deutsche Telekom both employ over 200,000 people. (Although that pales besides the 480,000 people employed by China's DGT, easily the world's biggest telephone company in terms of employment.) Overheads cost money. The International Telecommunication Union (ITU), an official club of the world's big telephone companies, puts the average cost of an international call at around $0.25 per minute. On some well-used routes, such as across the Atlantic, it is far less.

But those costs, remember, have rarely been tested by competition. The business of carrying international calls has been a monopoly—or near monopoly—almost everywhere. So has the business of delivering local telephone calls. Only as competition blossoms, internationally and locally, will it help to drive down all the other costs of a call.

The coming of competition

The big international telephone companies—including America's AT&T and MCI, Deutsche Telekom, and Britain's BT—have been good at controlling prices. How have they managed to do so? First, they have constituted a tight-knit but unofficial club, sensing—correctly—that it would be suicidal to compete head-on against one another. In the 1990s that club grew stronger, not weaker, thanks to a number of global alliances and mergers. The bulk of international telecommunication traffic is provided by just four big groups—AT&T-Unisource (linked to an alliance of several middle-sized carriers called WorldPartners), Sprint/Deutsche Telekom/France Telecom (known as Global One), Concert (the product of a merger, announced in 1996, between MCI and BT), and Cable & Wireless.

Second, the club owns most of the capacity that carries international calls. Building far more capacity than the market demands has given the club extra protection: telephone companies could always afford to drop their prices to kill off any upstart who tried to compete.

Third, the members of the club have been protected by their governments, which have allowed international calls to be handled either by a monopoly or by two or three companies: in the United States, AT&T, MCI, and Sprint; in Britain BT and Mercury, owned by Cable & Wireless. These lucky few have had the right to hand calls across the international gateway that separates one country's telephone network from another's. At the gate, predictably, they pocket a charge. This cozy arrangement has been backed up by a system of international accounting rates, which splits the cost of a call between countries. If one country puts through more calls to another than it receives, it hands over a settlement payment to even things up.

Now, the distance premium is being undermined by a number of factors: the end of government protection; the growth of new capacity built by companies outside the club; the resale of the telephone companies' spare capacity; the Internet; and, in the near future, mobile and satellite carriers. As Professor Eli Noam, of Columbia University in New York, observes, it's a buyers' market—there's no going back.

The end of protection

In some countries, governments have stepped in with measures that will help to drive down the prices of international calls. One step is to breach the monopoly at the gateway, by allowing many different companies to compete to carry international calls. Where true competition exists at each end of the call, as on routes between the United States and Britain, the effect on prices is striking. Such competition will be extended to many other parts of the world after the beginning of 1998, if countries stand by the promises they made at the World Trade Organization in spring 1997 to open their markets. The main countries of the European Union had previously agreed that they would allow telephone companies from other member countries to connect with their domestic networks on the same terms as domestic carriers.

In addition, the international accounting rate system is under attack. The United States is one of many developed countries that pays out more in settlements than it receives, running up a deficit of $5.1 billion in 1995. The U.S. regulator, the Federal Communications Commission, has been trying for many years to drive down the settlement rates that U.S. carriers negotiate with other countries. In early 1997, it threatened to do so unilaterally, by imposing "benchmarks." These are, in most cases, a tiny fraction of the bilaterally negotiated rates that are currently in force. If the threat plays out, it will have big repercussions for telephone companies in many small developing countries, which have relied on these payments for a large chunk of their revenue. But changes in the rules will have less impact on the busiest calling routes because more and more traffic on them has already begun to bypass the accounting-rate system.

Building new capacity

Building new long-distance capacity has become less expensive, allowing telephone industry outsiders to build their own links. In the mid-1990s, several new cables were being laid—mainly by newcomers to the international telephone business, entrepreneurs outside the club. The longest of these outsider cables, called FLAG (for Fiberoptic Link Around the Globe), runs from Britain to Japan, along the line of the Suez Canal and around the southern tip of India, and it will be able to handle 150,000 simultaneous calls. Nynex, an American regional Bell company, owns 40 percent of the project, but it has also attracted investors new to the international cable business, such as Marubeni, a Japanese trading company. Such companies recognize that transmission capacity will become a commodity—useful, scarce, and eventually as tradable as coffee or copper.[5]

Reselling spare capacity

The international editions of magazines such as *Time* and *Newsweek* are crammed with advertisements from companies promising huge discounts on international calls. Such companies—some of them owned by the telephone companies themselves—sell inexpensive telephone capacity to particular niche markets, such as travelers or tourists. Like the airlines, which sell off spare seats at low prices but try not to undermine their business-traveler market, the telephone companies sell their spare international capacity at different prices to different people.

Big companies and industries, such as the airlines, travel agencies, and banks, have long been able to lease telephone capacity to link their branches.[6] These private networks can bypass gateways and the complicated accounting-rate system. As private networks have multiplied, they, too, have begun to resell their spare capacity to other telephone customers.

That makes possible a variety of new services. For example, call-back services in high-tariff countries allow small or medium-sized business customers to make calls at lower rates. A caller in, say, Germany, telephones a number in a low-tariff country, say the United States. A computer identifies the caller without answering the telephone, rings back, and provides the subscriber with a dial tone with which to make a call to the United States or to a third country, say France. The call-back market is not huge in global terms, perhaps $600 million a year. But it is growing fast: in 1993 it accounted for sixty million minutes of traffic; in 1995, ten times that number.[7]

Country-direct and calling-card services, offered mainly by the American long-distance companies, constitute a much larger market.

These services allow travelers in a country where calls are expensive to telephone via their home operators and thus to be billed at lower rates. As a result, companies doing business in different countries can compete directly: for instance, someone calling from Britain can use Sprint rather than BT.

All told, these ways of cutting the costs of international calls (including call-back) probably amount to $3 billion to $5 billion of revenue, against a total of $55 billion for all international telephone calls. The market is not enormous, but it can have a big effect on some routes. In 1995, the demand for these services from India and from Hong Kong was about $300 million each.[8]

The Internet
The most important use for spare capacity on the telephone companies' networks is the Internet. On some routes, such as that between the United States and Japan, Internet capacity is already much greater than capacity for telephone voice conversations. But, paradoxically, voice conversations are an increasing share of Internet traffic: a number of companies have launched services that allow telephone calls to travel across the Internet. For example, in March 1997, USA Global Link, of Iowa, unveiled a service allowing callers in thirty-five countries to talk using touch-tone telephones for prices of less than $.50 per minute. Callers do not need a PC or an Internet subscription.[9]

Already, the Internet is a formidable rival to the fax, transmitting an international fax for the price of a local one. It has affected revenues on some routes, such as that between Australia and the United States, where (mainly because of the time difference) fax has in the past accounted for 60 to 70 percent of public traffic. But the question remains whether a large share of long-distance and international voice calls will eventually switch to the Internet. Even with only a tiny share of global telephone traffic, however, the Internet's effect on prices could still be huge. As Bill Gates, chairman of Microsoft, once said, "I'm not sure what the Internet is good for commercially, but I don't know why you would want to be in the long-distance market with that thing out there."[10]

As more and more big telephone companies offer access to the Internet, the tariff structures of the Internet and telephone networks will inevitably start to converge. The public Internet, however, will probably not become the main carrier of international voice calls for two reasons. First, the obvious customers for such a service would be small and mid-sized businesses. They may set more store by the quality of service that telephone companies provide than individuals would do. Besides, connecting a company telephone network to the Internet remains more complicated than connecting a single telephone.

Second, the public Internet simply does not offer enough capacity. As Vinton Cerf, one of the fathers of the Internet, puts it, "My back-of-the-envelope estimate is that if we took every bit of Internet capacity that the world has now and turned it over to Internet telephony, we might handle 3 percent of the world's load, and that's probably generous." Telephony absorbs a vast amount of bandwidth. For many countries, such às Australia, it would take quite a small growth in the number of Internet telephone calls to snarl up their entire Internet connection with the outside world.[11] But if Australians can talk to the world for the price of a local call, congestion is the predictable result.

Instead, telephone calls may increasingly be carried using the Internet standard or "protocol" but using networks separate from those used by public Internet. Telecom Finland offers one possible template. Beginning in mid-1997, Telecom Finland experimented with offering Internet telephony on corporate networks, allowing a company's workers at the click of a mouse to move from looking at a screen to talking to each other. Such services will become common and will exert a key pressure on conventional tariffs.

Mobile and satellite carriers

Eventually, mobile and satellite telephony will help to erode the distance premium. The first effect will be on long-distance charges. With a few exceptions (such as France), mobile telephone networks in most countries charge the same rate wherever a call is made—as long as it stays within national borders. At some point in the first decade of the next century, there will be enough mobile subscribers in many countries for the companies to compete head-on with fixed-line carriers on price. At present, many mobile operators are owned by fixed-line telephone companies. It will be the ones that are not—companies such as AirTouch, sold off by Pacific Telesis, a Californian Bell—that will lead the way.

Later, international call charges, too, will be affected. GTE Airfone represents one early sign of this change. It offers telephone calls from airplanes; in April 1996, the company launched a flat $15 fee for all calls, whatever their distance or duration. Eventually, that approach will be extended by the four separate projects—Iridium, ICO, GlobalStar, and Odyssey—now developing plans for rings of satellites that will allow callers to telephone from anywhere on the planet. Such calls, made over mobile satellite phones, may be priced between $1 and $3 per minute—less than the cost of an international call in many developing countries.

A new sort of pricing

In the future, pricing will show much greater fluidity and variability. Three main characteristics of the new pricing structure will be the end of cross-subsidies of local tariffs by the distance premium; a shift from charging per call to charging for the basic use of the network; and more frequent payment by the recipient rather than the person who makes a call.

Just as the airlines calculate that selling unused seats, however cheaply, is better than flying with those seats empty, so telephone companies will try to fill their empty networks by varying rates according to demand. As a result, just as no two passengers on an airline today seem to pay the same price for a seat, in the future no two callers will pay the same price for a call. Users will all get the same basic service (they all get on the same plane, or call the same town), but they will pay according to a whole range of special considerations.

In general, pricing will be based principally on the quality of service that a customer wants. Different tariffs, special offers, and discounts will proliferate. In the United States, this trend is already visible: the actual billed prices people pay for telephone calls are less than half the published price of a peak-rate call.[12] One guess at the future, by the ITU, is shown in Figure 2-2.

The end of the cross-subsidy

The end of the cross-subsidy of local calls by the distance premium will constitute a fundamental change in pricing. Generally speaking, the

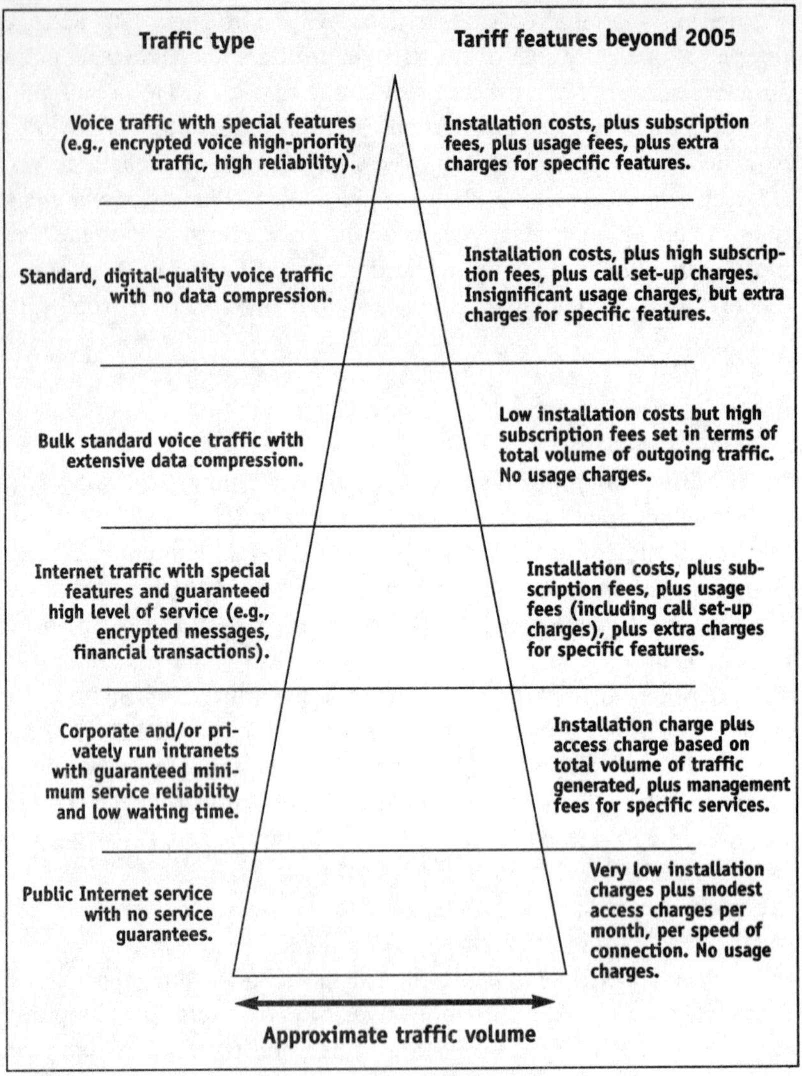

· · · · ·

Traffic type	Tariff features beyond 2005
Voice traffic with special features (e.g., encrypted voice high-priority traffic, high reliability).	Installation costs, plus subscription fees, plus usage fees, plus extra charges for specific features.
Standard, digital-quality voice traffic with no data compression.	Installation costs, plus high subscription fees, plus call set-up charges. Insignificant usage charges, but extra charges for specific features.
Bulk standard voice traffic with extensive data compression.	Low installation costs but high subscription fees set in terms of total volume of outgoing traffic. No usage charges.
Internet traffic with special features and guaranteed high level of service (e.g., encrypted messages, financial transactions).	Installation costs, plus subscription fees, plus usage fees (including call set-up charges), plus extra charges for specific features.
Corporate and/or privately run intranets with guaranteed minimum service reliability and low waiting time.	Installation charge plus access charge based on total volume of traffic generated, plus management fees for specific services.
Public Internet service with no service guarantees.	Very low installation charges plus modest access charges per month, per speed of connection. No usage charges.

Approximate traffic volume

Figure 2-2   How Pricing Might Look in the Future
A possible scenario for the evolution of pricing by different types of traffic.
Source: Adapted from International Telecommunication Union/TeleGeography Inc., "Direction of Traffic, 1996."

profits from long-distance and international business have helped telephone companies to hold down charges for local calls. International calls, in particular, have been the most profitable service offered by the large telephone companies. As the cross-subsidy disappears, the prices of local calls often rise. That can be political dynamite: in Italy in early 1996, trade unions and consumer groups made such a furious fuss about a plan to raise peak-hour local charges while cutting long-distance and international rates that the plan was shelved.

In the long run, though, such "rebalancing" will benefit telephone users. As long as the established companies cross-subsidize local service, new entrants will find it hard to offer them profitably.

A shift to up-front charging

At present, most telephone companies charge customers mainly by how much they use the network. In the future, access charges—an up-front payment for use of the network—will become the main source of telephone-company revenue. Customers may be offered "all-you-can-eat" charging, which many Americans already have for local calls. In that way, telephone tariffs will become more like the Internet, which is financed almost entirely with connection and subscription charges, rather than through payments for each use.

More calls paid for by the recipient

A growing share of calls will not be paid for by the caller. As individual calls become less expensive, companies will find it increasingly attractive to offer toll-free services to their clients and suppliers. In the United States, toll-free calls already account for 100 million calls per day, and AT&T reckons that some 40 percent of its domestic traffic travels to toll-free numbers.

A new sort of telephone industry

The changes in the economics of the telephone business will go hand-in-hand with changes in what that business looks like. Up to now, telephone companies in most countries appear to have largely controlled their own destinies. As technology and the regulatory framework change, that era will end.

In the late 1990s, the global telecommunications industry seemed to be standing where the international airline industry had been a decade and a half before. At the start of the 1980s, the big carriers were boom-ing. Ten years later, after privatization, deregulation, and cutthroat competition, many were in deep trouble. In the period between 1990 and 1992, according to some estimates, the airlines lost more money than they had made in all the first sixty years of their history.

The global telephone industry is at least twice as large as the airline industry, and it now faces greater upheavals than the airlines ever did. The airlines, at least, escaped the relentless onrush of technological change that is now transforming the telecommunications business. Spare airplanes have a value—and they can be moved from one part of an airline's network to another. No cost is as sunk as that of an unused, outdated telephone network.

On the other hand, where competition has been encouraged, airlines were becoming profitable again by the mid-1990s. Even where it has not, many national carriers such as Alitalia and Air France, which by rights should long ago have died, were still flying, protected by state controls (especially over landing slots) and state subsidies. Moral: when government pride is involved, change can often be slowed.

The telephone companies face this period of upheaval with several strengths: a lack of debt, for the most part; well-known brands (some 40 percent of Americans, according to one survey, still believe AT&T is their local telephone company, although it ceased to be in 1984); control of the main network with which all other telephone companies need to connect; and sunk costs that leave individual companies free for long periods to price their services below the level newer operators could afford to match and spreading the cost of billing systems, marketing, and developing software across large numbers of customers.

If the telephone companies are to survive in better shape than the airlines did, they will need to change much faster. Airline passengers might notice if their flight attendant did not speak their language, if the sandwiches were stale, or if their flight forced them to travel by a long, circuitous route. But telephone traffic is a commodity. If an Asian or Latin American carrier "hubs" traffic through a country on the other side of the world, a caller will probably never notice. Even more than with air travel, the price will be what matters most.

Beyond distance

The premium for distance will disappear, fairly quickly on some routes, such as the Atlantic, with lots of competition, more slowly on others, with weak competition. Declining rates will have powerful effects. As the ITU puts it, "It seems inevitable that a world without distance is also a world without borders."[13] Not only do some countries feel that their national identity is bound up with their communications monopolies—national post, national telephone giant, and so on—they also, more importantly, accept that calls should cost less within their borders than beyond them. The difference has helped to define "abroad" and "at home."

In Norway, with four million people in a country more than one thousand miles from end to end, long-distance telephone rates have been kept deliberately low for many years, to encourage a sense of nationhood. In the European Union in the mid-1990s, by contrast, it cost roughly three times as much to call Paris or Bonn from London during the working day as to call Penzance, a similar distance away. By using a calling card, it is often less expensive to call the United States from Europe than to call from one EU country to another. The death of distance in telephone pricing might not make people love their neighbors, but at least they would be more likely to know each other.

Versatility: New Tricks for Old Networks
. . . . .

It is easy to forget what a remarkable technology the telephone is. A single standard allows more than 700 million machines to connect with each other, carrying speech, data, and faxes to compatible machines from Michigan to Morocco. And the telephone keypad, with ten numerals, the alphabet, and a few extra symbols, can be used, at one extreme, to ring the house next door and at the other, to turn the telephone into something like a small computer.

For a familiar object that has been in use for more than a century, the telephone shows an extraordinary ability to evolve. In not much over a decade it has become mobile and truly interactive, in the sense that the user can send a return signal by pressing a key. Telephones are beginning to have screens, more elaborate keypads, and processors. As communications develop, so the telephone's ubiquity and simplicity will ensure that it develops many more uses. Why should customers have to learn to work an unfamiliar technology (and, even in the United States, only 40 percent of homes have a personal computer) when the familiar telephone can be taught new tricks?

Three main transformations are taking place in the telephone today. First, simple novelties are proliferating. As the capability of the telephone network increases (and especially as digital networks are installed), a large range of new services can be offered to customers without much change in the equipment they use. Second, wireless has transformed the telephone itself and has also created a huge range of new opportunities for services that track or locate moving objects. Third, the expanding transmission capacity of telephone networks allows completely new sorts of services to travel across telephone connections, most obviously in the form of the Internet.

For individuals, the telephone will remain the principal device for long-distance personal communication for at least the first two decades of the next century. Indeed, it will become even more personal, as telephone numbers become routinely portable, even when a user switches from one service provider to another. Eventually, numbers may be allocated to individuals, not to telephones in homes or offices. As a result, personal numbers may become a reliable way to reach someone even after five or ten years, even when their jobs and home addresses have altered. They will become part of people's identity, like their Social Security number or date of birth. Perhaps, too, personal numbers will be replaced by something more memorable and humane than a string of digits.

The telephone's ability to carry the human voice, uncomplicated by pictures, is part of its charm and its intimate power. A telephone conversation is an experience unlike any other that sighted people normally engage in. Its one-dimensional quality means that, like talkers in the dark, people on the telephone may feel able to say things to each other that they would find it difficult to say face to face: to sell, to threaten, to seduce, to lie. Invisibility liberates: you can chat to your boss or your mother while glancing through the newspaper or getting dressed. But the camouflage is only partial. Undistracted by body language, which may deceive the eye, the ear often picks up faint emotional signals—distress, glee, equivocation. The telephone sharpens the senses, allowing the ear to spy on the voice.

Simple novelties

Many of the new powers currently being acquired by this intimate instrument are embedded in the network to which it is connected. This model differs from that of the Internet, where the intelligence resides not in the network but in the computers, large and small, at its extremities. The customer's reliance on the telephone company to provide new services through its network gives the companies a powerful competitive tool. But it also means that customers may find it harder to shop around for the precise combination of services that fit their individual requirements.

Many of these new services are activated by pressing buttons on the keypad of a touch-tone telephone connected to a digital exchange. Such telephones are still relatively rare around the globe: by some accounts, perhaps 80 percent of all telephones are still of the old sort that operate on pulses, not tones.[14] But, where it is possible, button-pressing effectively gives the telephone a new sort of interactivity: the ability to send messages. Given a digital exchange, it is now possible to pay a credit-card bill or check the balance in a bank account entirely by pressing buttons on a touch-tone telephone in an appropriate sequence. But, in fact, while that option may save costs, it may also lose customers in a maze of recorded messages ("press eight for yet another service you do not need"). Whatever clever services the telephone offers, people will long continue to want a real human being at the other end.

Many new services entail fancier ways to send messages. Already, voice messages can be sent with a time delay or to a group of people on a distribution list. In the future, unified message systems will allow people to retrieve voice messages, faxes, and electronic mail from a PC or a touch-tone telephone. Recipients will be able to print out messages or have them read out (electronically) if, say, they want to listen while they drive.

. . . . .

Speech and voice recognition will become a reality. For years, speech recognition has been promising much and delivering little. Now, three factors have given a fillip to telephones that respond to simple spoken commands: the need to be able to use telephones safely while driving a car (to dictate a password, say, rather than punching in a string of numbers); the need to be able to give commands (for example, to get into a message mailbox) when a touch-tone telephone is unavailable; and the need for security.

Telephone users will increasingly be offered ways to forward telephone calls to wherever they happen to be in the world. In one sense, this is not an innovation: call-forwarding is old hat in the United States. Previously, such facilities were available only to people who worked in the same building or in the same local-exchange area. But now, networks can offer personal numbers that allow an individual to be tracked down almost anywhere in the world. Increasingly, the telephone is becoming a tool to contact a person rather than a place.

Telephones can now identify callers, even if the call is not answered, something of particular importance for telephone commerce. It has made possible the development of call-back facilities, and it allows companies to collect information on would-be customers and to check the efficiency with which staff answers calls.

Two other familiar services will continue to develop. In spite of competition from electronic mail, the fax will continue to have a separate existence for some time, partly because verifying receipt is easier with a fax than an e-mail, but also because a fax can transmit a signature. Gradually, the fax and the Internet will converge: increasingly they are used interchangeably, with faxes sent to computers and computers used to send faxes.

The pager, too, will continue to evolve. To send textual messages to a telephone is easy and inexpensive: unlike colored pictures or moving images, text requires little bandwidth. The most fanatical users of pagers are in Asia. Japanese schoolgirls jam the networks of the Japanese telephone system at bedtime each evening with messages wishing their girlfriends sweet dreams. In China, a coding system has been developed that allows pagers to relay long messages.

It is still not clear how all the telephone's new and fairly new tricks will be used. Although most are commonplace in the United States, in many countries they are only beginning to become available. A survey by Dataquest at the end of 1996 on voice messaging in Europe (a commonplace service in the United States) found that it was used by only 35 percent of British companies, by a mere 18 percent of German firms, and by a paltry 15 percent of French ones.[15]

The most important novelties, though, may turn out to be those that allow the Internet to be integrated into the telephone service. Many companies now make it possible for callers to type personal details onto a Web-page application form, which then goes straight to a corporate sales department—to apply for a home loan, say, or to take out a subscription. But the more important innovation allows a caller to click on a screen and be put through to an operative—at a bank, say—and to talk with that person while both look at the same page of information (account details, say) visible on their individual computer screens. In such integration of Internet and telephone lies the future.

Wireless telecommunication

By far the most striking change in telecommunications since the early 1980s has been the development of wireless services. It has brought innovations of three kinds: the mobile telephone; the wireless link to a fixed point, known as "wireless local loop"; and the tracking of moving objects.

Mobile telephones

In the space of little more than a decade, more than one in seven of the world's telephone subscriptions are for mobile telephones. By the turn of the century, a quarter of the world's 1.2 billion telephone subscribers will have mobile telephones.[16] As the use of mobile telephones continues to grow, there will be three key trends: mobile telephones will capture a growing share of voice traffic; the line between a mobile and a fixed telephone will blur; and mobile telephones will acquire many of the attributes of portable computers.

At present, almost all of the world's 135 million cellular subscribers[17] also have a fixed-wire telephone. Fixed and mobile services coexist and are often owned by the same big company. (AT&T, for example, owns McCaw Cellular, the world's biggest mobile operator.) Since virtually all calls to or from mobile telephones currently finish or start on the main operator's fixed network, the mobile telephone is a lucrative source of new business for big telephone companies.

But new mobile services have sprung up, partly because governments have been more willing to allow competition here than in fixed-line services. The mobile telephone is now the main competitor for the telephone companies' core business of voice conversations. In time, most voice conversations will be received on and sent from wireless telephones (although in between, the connecting string may still be a fixed, not a wireless link). In countries that are building telephone networks from scratch, that point will come quickly: for example, Cambodia already has more mobile than fixed-line telephones.

The trend is apparent in Sweden, where fixed-line connections declined in the mid-1990s because young people do not want to incur the high cost of being connected to the fixed network (or the nuisance of changing their telephone number) each time they change jobs, partners, or apartments. By the end of the century, there will be many more Swedens.

The shift from fixed to mobile telephony will be hastened by the falling price of actually making calls over mobile telephones. In 1992, mobile service tariffs were, on average, three and a half times higher than fixed-link tariffs.[18] That rate has now fallen to a point where the overall cost of using some mobile services is becoming competitive with that of fixed services. Once this becomes general, a mass market will have been reached, and voice calls from mobile services will start to overtake those from fixed services.

The fusion of fixed and mobile telephones will take place on many fronts. Mobile telephones will become available that are not tied to one service (and tariff), for example, and handsets will act both as mobile phones and as conventional fixed links. The handsets of mobile telephones will increasingly incorporate several different receivers. A handset that can use the world's main cellular standards could be used by business travelers around the world. A telephone that can switch to the most efficient transmission system—a cordless telephone in the home, a cellular telephone in a fast-moving automobile, a satellite phone in a remote location, and so on—will use radio spectrum more frugally.

"Personal communications services," a new generation of wireless telephones, will be combined in all sorts of flexible ways with cordless and fixed-wire telephones. Japan's "personal handy-phone system"

(PHS), launched in 1995, provides an example of such "dual-mode" phones. Pick it up to go into the yard, and it becomes a cordless; walk down the street, and it locks onto the beam from a local aerial. This tiny device, the size of a powder compact, offers four hundred hours of battery life on standby and five hours of talking time. It works through small, low-powered base stations sited close together in big cities. Its airtime charges are set to compete with those of pagers: they are only one-fifth those of cellular systems, and local calls cost less than calls from pay phones. Its aim is to turn the mobile telephone into the standard means of communication for ordinary people. Salesmen tout it to students outside school gates and to housewives in supermarkets. By 1997, it had become cheaper to use a PHS in the home than a fixed-line phone. The main drawback to PHS is that it cannot be used when traveling at speed (a luxury that Japanese drivers rarely enjoy). Other companies, such as Motorola in the United States and Seimens in Europe, are expanding the concept, so that eventually it will be possible to use the phone while driving.

Many people will carry mobile telephones as unthinkingly as they carry a wallet or a watch. Mobile telephones will probably be built into many items: they might become a standard attachment for a handbag or briefcase, perhaps, and certainly in a car. They may become even smaller (the limiting factor may be the size of the keypad). Their battery life will lengthen. They may be built into increasing numbers of devices, such as the earphones of a portable radio or a hearing aid—when, with developments in voice recognition, it becomes possible to do away with a keypad.

Mobility and computing will become increasingly linked as devices for harnessing mobility and the power of the computer, such as Hewlett-Packard's Omnigo or Nokia's 9000, continue to evolve. Such gadgets can already carry out tasks as varied as searching databases, sending e-mail, and taking orders in restaurants. Some of them will be able to read handwriting, although computers seem have found it every bit as difficult to learn to do that as to recognize spoken words. Up to now, attempts to develop a populist version of a "personal digital assistant," such as Apple's Newton, have failed to take off. Specialist uses are likely to grow faster than mass-market ones until handwriting or speech recognition improves.

The booming global demand for pagers suggests huge potential for inexpensive mobile data communications—especially when combined with a service. Asia's army of small businesses frequently use secretarial paging services. A customer calling a company's number is answered by a secretarial bureau; a secretary pages the boss, who picks up a pay phone to return the customer's call. The customer has no way of knowing whether the company is housed in a palatial office suite at the top of a downtown tower or in a back-street shed.

Wireless local loop

Building a local telephone network is expensive. "Wireless local loop" offers a cheap alternative. A small fixed radio antenna installed in a house or office receives and transmits calls through a nearby transmitter. While a mobile telephone uses a great deal of spectrum to switch from one base station to another, a fixed antenna can be permanently tuned precisely to the correct base station.

Because this system relies on sending signals through the air, it is intrinsically less expensive than a purpose-built fixed link. No need to dig up roads or to pass the homes or work places of people who do not want to be connected to the system. The customer comes to the telephone seller, not the other way around. The main cost is that of transmission: spectrum must be allocated. But maintenance costs are low. Most faults arise at one end or the other of the network, making them easier to repair than those in the part between exchange and customer.

Such links therefore offer a quick and inexpensive way for new entrants to develop telephone services. So far, they have been used mainly to build telephone systems quickly in developing countries and in Eastern Europe and the former Soviet Union. Fixed wireless is especially useful for cutting the cost of connecting rural communities. A few examples of successful installations provide a sense of the potential of this technology. Starlight Communications has installed wireless public telephones in Somalia; the Republic of Tatarstan is building a large-scale telephone network based entirely on wireless; and in Ghana, Capital Telecom is installing a wireless local-loop system to serve fifty thousand rural subscribers.[19]

As competition moves into local networks in rich countries, some companies are building wireless local loops there, including Hughes in the United States and Ionica, a subsidiary of Canada's Northern Telecom, which launched a wireless telephone service in Britain in May 1996. Some analysts argue that wireless can be used to build a local-telephone network from scratch for perhaps half the cost of a fixed-link network.[20] But there has been a catch: bandwidth limits have prevented wireless services from providing high-quality Internet and video connections. In parts of the world where no fixed-wire telephone service exists, that is not a disadvantage. In the rich world, however, it is. Now that may change. In February 1997, AT&T decided to develop a fixed wireless system to bypass local telephone communities. The company claimed that it had found a way to carry far more data than can be squeezed through copper wires, and thereby to make high-speed video conferencing and Internet access available.[21] If so, that would eventually overcome one of the main drawbacks of wireless.

Tracking moving objects
Mobile communications have another set of applications just as radical as the mobile telephone: locating moving objects. With the help of global positioning satellites, a position on the globe can be pinpointed with extraordinary precision and, with land-based wireless, moving objects can be tracked.

The technology received a boost during the Gulf War. The American army, finding that it could not know without going and looking at them what was in the containers being shipped out to the desert, installed a system, developed by Savi-Technology (a California company now owned by Texas Instruments), based on two-way battery-powered radio tags. The tags stored information on the containers' contents, destination, and schedule and could be tracked by satellite.[22]

From such technologies, a huge range of uses is developing, including the following:

- Inventory control systems: systems used to monitor the position of electronically tagged containers, for instance, or to check stock on supermarket shelves;

- Navigation systems: electronic maps that pinpoint the location of an automobile, truck, or military target or that help a hiker find a track or a farmer to work out which parts of a field to spray;

- Alerting systems: electronic tags that track stolen automobiles or roaming elephants or probes that monitor the state of a patient's heart to warn of an impending attack;

- Toll payment systems: systems for automatically assessing tolls against automobiles passing through an electronic checkpoint.

One sort of tag is passive. It contains a chip and antenna but no battery; when triggered by a radio signal directed toward the tag, the chip transmits an identity number. Another more powerful, but less long-lasting chip, such as those on the Gulf War army containers, contains a battery, a radio receiver, and a transmitter.

The most common uses for all these radio-tracking technologies is in transport. Automobiles, increasingly, are riddled with tiny computers. With the addition of a mobile link, they could be used to tune an engine or do spot maintenance, in effect servicing the car by phone. Other car-borne devices could be used to direct drivers to less congested routes or to track stolen vehicles—or, perhaps, to charge a motorist a premium for driving down certain congested or ecologically sensitive streets. Fuel pumps at some Mobil petrol stations in Missouri have been equipped with tag readers so that customers can be billed automatically when they refuel their cars.

Living with the Internet

When the telephone companies belatedly noticed the emergence of the Internet and the World Wide Web, its multimedia side, they found that a completely new use for a telephone connection had been growing at explosive speed, not because the telephone companies promoted it, but in spite of their hostility or apathy. It is by far the most dramatic change in the business of telecommunications in the late 1990s, and both a threat and a boon. It puts pressure on telephone networks and pricing systems. But it will also create new opportunities. The growth of the Internet gives the telephone companies a chance to carve out a new role and provide a new service.

Pricing problems
People who use the Internet from home (or from a small business) generally use an ordinary telephone connection. The Internet therefore boosts the lucrative demand for second telephone lines: in 1996,

Pacific Bell installed a record 700,000 telephone lines in California, more than twice as many as in 1994. Indeed, according to one study, America's regional telephone companies earned more than $3.6 billion in revenues between 1990 and 1995 from additional lines used exclusively or mainly for on-line services.[23] For telephone companies in countries where calls are charged by duration, the Internet has dramatically increased new call revenue. In Britain, for instance, on-line connections accounted for about 10 percent of all local-call minutes carried by BT at the start of 1997.

Where calls are not time-charged, growth has been even faster. In the United States, the average call to an Internet service provider lasts from two to four times as long as an ordinary residential call. Where Internet use is heaviest, the effect is to clog the local network, which was designed for short chats, not long surfs. In California, according to an early 1997 study on the Pacific Bell network, 27 percent of all calls from homes were Internet calls. By 2001, Internet calls from homes on the network will almost equal voice calls from homes.[24]

As a result, telephone companies in the two countries where time-charging is least common, the United States and Australia, want some kind of length-of-call charge for local calls. In March 1997, the Australian government debated a bill to allow Telstra, the main telephone company, to time-charge local calls to the Internet while continuing to exempt ordinary voice calls. This is just one example of the quandaries that the Internet creates for telephone pricing policy. Ironically, while some telephone companies agitate for the introduction of time charges, others undermine them by becoming Internet service providers and frequently offering subscribers several hours of Internet access for a fixed sum.

The telephone companies have still not worked out how they can benefit most from the Internet. The most adventurous, like Telecom Finland, are experimenting with providing an Internet telephone service. Others, like MCI, are building their own high-quality Internet on which they will provide a premium service for customers (such as banks) that want secure and uncongested connections. Still others wonder whether they can become the main gateway between the customer and the electronic marketplace, supplying traders with information on their customers and billing shoppers on the traders' behalf. One thing is sure: if the telephone companies are to provide services that are more valuable than the commodity business of carrying calls, they will face fierce competition from the myriad small companies with low costs and no baggage of entrenched ideas that now provide the bright ideas about the future of the Internet.

Local capacity

If the telephone companies are to develop the quality Internet access their customers want, they will have to invest in increased capacity for their local networks. It may seem odd to worry about the capacity of the telephone network after all the discussion above of glut in parts of the international network. But, while a single strand of glass can carry nearly all of Europe's transatlantic voice traffic, the connection to the front door remains a pair of twisted copper wires, designed for nothing more than two human voices. The resulting bottleneck outside the customer's front door has become an increasing problem as more people want to use the Internet from home.

Running fiber-optic cable to the customer's door is one way to increase capacity, and that is already happening in many places. Fiber connections run into many center-city businesses and into apartment blocks in dense cities such as Hong Kong. In Italy, Telecom Italia has a project to lay fiber to its customers' homes, probably to stop them from defecting to future competitors. U.S. telephone companies are introducing fiber into the outer parts of their residential networks as part of routine maintenance, at a rate of about 5 percent of total wires per year.[25] Sound commercial reasons dictate this. Broadly speaking, maintenance accounts for about one-quarter of the costs of running a network, and the maintenance costs of a fiber-optic network are around one-fifth those of a wired one.

On the other hand, while the cost of a new long-distance connection is spread across millions of users, the cost of the last part of the link is met by a only handful of households or a single business. Digging up roads is extremely expensive; and plenty of homes may not want or be able to pay for the extra services that then become available. So the telephone companies are keen to develop speedier ways to increase the carrying capacity of their copper. Another way to—as it were—squeeze elephants through the drainpipe of the local telephone wires is to develop techniques of digital compression.

Table 2-2   Representative Costs of Local-Loop Technologies

Component	Cost per Connection ($)
Copper	200–2,000
Fixed wireless	500
Coaxial cable*	500
Fiber to curb	1,050 upward
Fiber to home	2,000 upward

*Does not include additional cost to upgrade for interactive capability.
Source: *Information Infrastructures: Regulatory Requirements.* ©OECD, 1997, Paris. Reproduced by permission of the OECD.

One such technology, Integrated Services Digital Network, or ISDN, is widespread in Japan and continental Europe but has been little used in the United States since it was developed in 1982. Another, more recent and with one hundred times the capacity is Asymmetrical Digital Subscriber Line, or ADSL, and its technological cousins, known collectively as XDSLs, all of which increase the carrying capacity of copper sufficiently to allow it to carry video services. But both are expensive and cumbersome, and most customers are not yet willing to pay. As of 1996, ADSL still cost between $1,500 and $3,000 per home.

Overall, the relative costs of the various options for connecting local subscribers vary enormously. Table 2-2 shows one range of estimates. These involve a fair number of apples-to-pears comparisons—optical fiber to the home would carry far more information than would wireless, for example—but they give a rough idea of orders of magnitude.

As the telephone companies gradually increase the capacity in their local networks, they may well also become the main providers of Internet access. They will use the Internet to improve their own services, blending together the telephone and the computer. Their services, like their pricing, will thus merge seamlessly with those of the Internet.

Wiring the World

. . . . .

While telephone companies in the rich world struggle to think of new and more lucrative services to offer their sated customers, two out of three people on the planet still have no access at all to the telephone. But the good news is that, in many parts of the world, people are receiving telephone service for the first time. New telephone lines are being added almost twice as fast in the industrializing countries as in the industrialized. The result is an immense boom in new investment. Some analysts estimate that more money will be invested in telecommunications networks in the 1990s than in all the previous years since Alexander Graham Bell patented the telephone.[26]

Some developing countries are installing telephone connections at an astonishing rate. Asia has had the world's fastest growth in telecommunications, adding new lines at an annual rate of about twenty to twenty-five million, or three times as many as are added in the United States each year. China alone adds about fourteen million lines per year and intends to increase that to twenty million per year by the turn of the century. That will be the equivalent of duplicating the line capacity of America's Bell Atlantic every year. Lower-middle-income countries—such as the Philippines and Venezuela—see rates of growth for newer technologies such as cellular telephones that are double the rates in high-income countries.[27]

But in other parts of the developing world, progress is slower, and demand vastly exceeds supply, even at the exorbitant tariffs that are often the norm. Officially, some 43.4 million people around the world are waiting for a telephone line, with a typical wait being about fourteen months.[28] In reality, the number of potential customers waiting for service is probably far higher: those who see no hope of a connection do not even bother to register.

In building their telecommunications networks, poor countries have two advantages: many new technologies reduce the cost of installing new networks; and their newly installed networks will be state-of-the-art, unlike those in the rich countries, which now require upgrading.

Cost-saving technologies can reduce the cost and increase the reliability of services. One possibility is wireless local loop, discussed above; another is pagers, also discussed above, which many Asian countries use as a first step to full-blown telephony; a third, useful for cities in poor countries, may be a version of Japan's PHS. Another option is "virtual telephony," a technology that provides people access from any telephone to a secure electronic mailbox.

Developing countries also have the opportunity to go straight to entirely digital networks. For instance, countries with all-digital networks include the Maldives, the Solomon Islands, Djibouti, and Rwanda. By contrast, in 1995, only 56 percent of Germany's network and 73 percent of the U.S. network was digital.

Developing countries can also move to mobile telephones more rapidly than can rich countries. Their mobile operators do not need to compete against a massive installed base of fixed-wire telephones, which has long since been amortized, and their spectrum is frequently less heavily used than in the rich world. The speed with which a wireless connection can be set up—a matter of hours—is in stark contrast to those interminable waiting lists for fixed-line telephones.

Privatization, competition, and open markets assume even more importance in developing countries than in rich ones, because their national monopolies so often suffer from a mixture of lack of capital and bad management. In fact, developing countries have some of the most radical examples of market-opening, often with correspondingly dramatic improvements in service. In Chile, the first Latin American company to privatize, the proportion of people with access to a telephone doubled within three years and waiting time fell from nearly ten years to just one.[29] Peru sold a 35 percent stake in its telephone system to Telefonica de España in 1994, a rare instance of a country ceding effective control of its system in exchange for a large investment in it. Several developing countries have accepted bids from foreign firms to build parts of their networks: thus USWest is building a network in southern India, AT&T is building a mobile network in Argentina, Deutsche Telekom is building a fixed-link system in Indonesia, and Australia's Telstra is installing pay phones in Phnom Penh.

Such policies make sense not only because they will allow users access to inexpensive telephone services, but for an even more important reason. Good telephone services are especially important for poor countries because they bind the poorest, most distant regions into the rest of the country, and they allow developing countries to compete on the same footing as richer rivals.

Benefits of telephone service to rural areas

People who live in the country are less than half as likely to have access to a telephone as people who live in cities, a perennial problem in poor countries. Sometimes the discrepancy is much worse: in India, 74 percent of the population live in villages, but they have only 10 percent of the lines.[30] This is partly because city dwellers have more political clout than rural dwellers and partly because it costs up to five times as much to install a fixed line in a rural area as in a town. An obvious answer is to install wireless links—mobile or, even less expensively, fixed.

In rural areas, good communications can bring news, education, and medical and agricultural advice. They also provide a disincentive to move to the city: why uproot, if some of the advantages of urban life are available to those who stay put? For poor countries, where most of the world's megacities have sprung up, the benefits of persuading people to stay in the country are immense. Moreover, rural telephone services frequently earn much larger revenues than those in cities: country dwellers make—and attract—many more long-distance calls than do city people.

It is every bit as important to connect cities to the rural areas in their hinterlands as it is to connect cities to the rest of the world. Failure to do so will mean that the impact of the death of distance on poor countries will be divisive, not unifying.

Benefits of telephone service to competitiveness

One of the powerful effects of the enormous investment in telecommunications now under way will be to reduce some of the gaps between developed and developing countries. Since around 1990, international-call charges in the two have been converging as developing countries cut tariffs at a faster rate than rich countries (by roughly 3 percent per year, to the rich countries' 2 percent). The fast-growing developing countries, especially in Asia, are now on track to become the telecommunications powerhouses of the next century. Already, today's emerging economies account for more than 25 percent of global telephone traffic, and by 2005, the ITU predicts that will rise to more than 50 percent. By then, China will be one of the top three sources of international traffic, and India will be close behind.

The benefits from plugging developing countries into the vast global telephone network will be enormous. One day, developing countries may be a low-cost location for telecommunications services, just as they now provide a low-cost location for producing textiles and electronic goods. Indeed, this trend has started, with, for example, call centers (switchboards that handle many calls to a single number) being located in the Philippines. By the start of the next century, rich countries may find it less expensive to reverse the direction of a call to exploit low telecommunications costs in the developing world than to use callback services located in places like Seattle and Florida.

In addition, the telephone brings access to information from other parts of the world: technological information, which can allow developing countries to catch up with the rest of the world in industry, health care, education, and a host of other services; commercial information, raising aspirations and transforming patterns of consumption; and political information, which will foster the spread of democracy and open discussion of public affairs. As investments go, wiring the world is a formidably productive one.

Inexpensive telecommunications services will narrow the advantage of the first world over the developing countries. In one sense, indeed, countries with good communications will be indistinguishable. They will have access to services of first-world quality: health care, education, technical information. They will be able to join a world club of traders, electronically linked, and to operate as though geography had no meaning. This equality of access will be one of the great prizes of the death of distance.

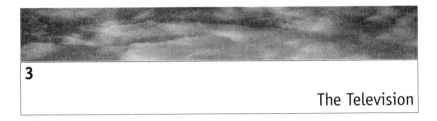

3

The Television

Television, predicted the *New York Times* on the occasion of the 1939 World's Fair, "will never be a serious competitor for radio because people must sit and keep their eyes glued on a screen; the average American family hasn't time for it."[1] Even journalists sometimes make mistakes. Midway through any given evening in Europe or the United States, about half the people are doing the same thing: watching television. This activity now takes up more of the average person's time than anything else, apart from sleeping and working. People in Western countries typically spend between one-half and one-third of their leisure time in front of the set. The ability of television to entertain, inform, titillate, and lure has made it not only the chief leisure activity of the twentieth century but also perhaps the most influential cultural invention since the printing press.

Now, though, technology is transforming television as surely as it is the telephone. Two enormous changes are under way: a vast rise in the capacity of delivery systems, so that people can receive not hundreds but thousands of different channels, and a fall in the cost of making and distributing programs. These changes will alter the nature of television. A third change is just beginning: the convergence of television with the Internet.

But the changes will differ in important ways from those that are bringing the death of distance to the telephone. First, most of what people watch is "free," so the fall in delivery costs now cutting the price of a telephone call will not cut the cost of watching television. Instead, ironically, people may be asked to pay for the programs they most want to watch. More and more television will be sold, as cable and satellite television frequently already is.

· · · · ·

Table 3-1   Who Is Tuned In?

Income Groups	Receivers per 100 Inhabitants	Television Homes as % of Total Homes	Multichannel Homes as % of TV Homes
Low-income countries	**12.9**	**47.0**	**18.0**
China	24.7	62.5	15.3
India	6.1	31.7	32.3
Lower-middle income countries	**20.4**	**71.0**	**12.7**
Philippines	10.0	56.7	4.0
Poland	12.6	91.9	37.5
Upper-middle income countries	**26.0**	**89.6**	**44.7**
Brazil	27.8	87.7	NA
South Korea	32.1	99.3	66.4
High-income countries	**61.2**	**90.1**	**51.9**
United States	77.6	94.9	69.2
Singapore	36.2	88.2	5.9
World	**22.8**	**65.9**	**33.5**

Source: Data from International Telecommunication Union, "World Telecommunication Development Report 1996/97."

Second, in many wealthy countries, people already seem to be watching as much television as they want to: indeed, in some countries, viewing time is declining. In the future, the television will have to compete for viewers with other uses of the screen—such as games, socializing, and gathering information.[2] So, while time on the telephone (another competitor with television) is likely to rise, time watching television may actually fall. Television will offer far more to watch, but viewers will have less time to watch it.

In the developing countries, however, where channel choice may be limited and transmission hours short, the revolution will be different. For levels of television ownership are still far below those in the rich world (see Table 3-1). The rise in set ownership and the spread of multichannel television to these parts of the world represent a vast new market. Some of it will be fed by the exports of the rich world; much more will be fulfilled by the development of new local entertainment industries.

As channels multiply, the most basic question will be whether television becomes fragmented. Television has been a uniquely unifying national phenomenon. Never before have so many people held in common a core of shared cultural experiences (although culture may not be quite the right term for "The Oprah Winfrey Show" or "Wheel of Fortune"). That shared experience constitutes a durable communal bond. You may not know the names of your next-door neighbors, but you can be fairly sure that over the past few days they have watched some of the same programs that you have.

Now, with the vast expansion of television programming, everyone will be able to watch something different—"Me TV," perhaps—just as each Internet user can explore a different selection of sites. The television will become a personal piece of equipment—as is happening already in the United States, with four receivers for every five people—more like the mobile telephone than a communal source of entertainment. But it is also possible that, on these personalized machines, more people will watch fewer programs: that television will become more like the movie business, with a few blockbusters attracting vast global audiences.

Either way, people will have more viewing choice. In TV-saturated countries such as the United States, where multichannel television is commonplace, that will be important. In some parts of the world, it will be revolutionary. Large tracts of the developing world, including most of Africa, watch a few mind-numbingly boring and amateurish state-controlled channels.[3] Even in Europe, most viewers have little choice in programming. Austria, admittedly an extreme example, drafted a law allowing private television stations only in February 1997.

Viewers in all these countries will one day be able to pick their programs in a global market. They may still choose to watch national fare: imports, with the partial exception of American programs, tend to have smaller audiences than do national products. But, armed with a credit card and a remote control, people will eventually order television programs from anywhere they choose. The television business will then become truly global. So, perhaps, will the cultural values it instills.

Three key changes will take place. First, distribution networks will switch from analogue to digital and new delivery systems, especially satellite, will be developed. Second, content will increasingly be paid for directly by viewers. And third, the industry will increasingly be dominated by global alliances. Technological change will also bring a convergence of television and the Internet. The most important consequence may be to stimulate the same ferment of experiment and innovation in television that is currently taking place on the Internet. Combined with increased capacity in delivery systems, the Internet will enhance and extend the role of television, creating new applications and possibilities, not so much for home entertainment, perhaps, but certainly for corporate communications.

Changes in the Distribution Network
· · · · ·

Two big changes are taking place in the ways television is delivered. First, the old clique of over-the-air networks is being challenged by new delivery systems, such as cable and direct satellite broadcasting. Second, all delivery systems are becoming digital, thus hugely increasing their capacity.

The effect is to create, at breakneck speed, far more choice than has ever before been available on television. A medium that has long had limited capacity is thus acquiring more or less unlimited capacity.

More ways to distribute television

In most parts of the world, over-the-air television broadcasts have long been the main, even the only, sort of television available. But spectrum is limited, and analogue television needs lots of it. So the predominance of over-the-air (or "terrestrial") broadcasting has effectively restricted most people in most countries to four or five channels, often available nationally.

This restriction of choice has had enormously important consequences for the nature of television. First, because competition has been limited by spectrum shortage, television has been highly regulated, often even owned by government. Second, channels have had an incentive to cater to the largest possible audience, making television the ultimate mass medium. And, third, television's huge audiences have shaped the development of mass-market advertising, and so, for instance, the development of brands.

While the number of analogue over-the-air channels is tightly limited, most cable-television systems can deliver up to about fifty analogue channels.[4] In only a few countries, however, is cable television common.

The United States has easily the world's most extensive cable network, passing 80 percent of the country's 115 million homes and connecting 64 percent of the homes passed.[5] Only in a few European countries— Belgium, the Netherlands, and Switzerland—does cable run past a higher proportion of homes, mainly because these three small countries are surrounded by large neighbors with powerful transmitters that make broadcasting difficult. All told, the United States accounts for more than half the rich world's cable-television subscribers—for a good reason. The United States has a terrestrial transmission network of unusually poor quality; in most wealthy countries, tremendous public investment in transmission has meant that subscribers do not have this incentive to pay for cable delivery.[6]

Now, many countries have begun to build or to upgrade cable-television networks. Britain started to build a modern cable network only in 1984: by the start of 1997, 1.9 million homes, out of 8.4 million passed, subscribed to cable television. Cable-television services have also been growing rapidly in developing countries. In cities with many high-rise buildings, they offer a relatively inexpensive way to take television to large numbers of customers. By 1996, China already had more cable-television subscribers than any country other than the United States, and India had more homes with cable television than with telephones.[7]

But the biggest revolution in delivering television came in the late 1980s, with the development of what is termed in the United States "direct broadcasting by satellite," or DBS, and in Europe "direct-to-home," or DTH. Initially, satellites were used for telecommunications. Then, they were used to send programming to the cable companies, which would receive them at the cable "head end" and deliver them to subscribers. The big novelty of the 1990s has been the development of satellite television delivered straight to the customer who buys or rents a small receiving dish.

The fastest growth of DTH satellite in the early 1990s took place in Britain. Two rival companies—British Satellite Broadcasting and Sky Television—merged to form BSkyB, a company 40-percent owned by Rupert Murdoch's News International. BSkyB has been an astonishing success: within five years, it became Europe's largest media group in terms of market capitalization.

Lots of other satellite-television services have followed: indeed, by the mid-1990s the use of satellites for broadcasting was more wide-

. . . . .

spread and growing faster than its use for telecommunications.[8] Much of the growth was in regions—such as Eastern Europe and Latin America—where television has been relatively underdeveloped. In the United States, where DBS television began in 1994, it has been a runaway success: the digital satellite receivers for DirecTV, one of the pioneers, were the fastest-selling consumer electronics item in American history.[9] Unlike most countries, the United States has never had a single national transmission network. Instead, channels have been distributed by local television stations (often independently owned). Satellite has been able to offer a truly national service while catering for regional taste (in, say, sport).

By the middle of 1997, America's DBS services had attracted more than five million subscribers. That may not sound like much out of nearly 100 million television-watching households, but this rate of growth far outstrips that of the more established cable networks, which are admittedly more mature markets.

The boon of satellite delivery is its low start-up cost. The situation is the same as in the telephone business, where fixed-wire networks have first to be built past every home and business while wireless telephones go only to those who want them. In television, cable companies have to pay all the initial costs of building their networks, while direct-to-home broadcasters merely rent capacity on a satellite. In addition, while cable subscribers assume the cable company will carry the cost of connecting them, satellite subscribers usually buy the satellite dish and pay to have it installed.

In time, the telephone network may also become a delivery channel. In some parts of the world, the telephone companies are working on plans to put television services of their own through their networks, although the enthusiasm of the early 1990s shown for such services by most American regional Bells has since waned, a result of their new ability to compete for long-distance telephone traffic. Only USWest has pushed on with buying and building cable networks. In the communications business, as in most others, companies are generally better advised to do more of what they already understand than to throw money at a business they know nothing about.

The bonus for viewers of many delivery systems should be more choice: not just choice of content, but also, more importantly, choice in how to receive that content. Cable companies almost everywhere are local monopolies. Just as it is clearly desirable that telephone subscribers should be able to choose their company, so viewers should be able to choose among several companies that can deliver television to their homes.

Going digital

The start of digital television is the first big change in the way a signal is delivered since the launch of color. Digital compression allows many more digitally encoded channels—already between four and sixteen—to be squeezed into the space required for just one analogue channel.[10] As compression gets even better, the number will rise. Through digital delivery, televisions can use many of the tricks that were once confined to computers: storing information or programs, for example, and manipulating it in various ways. Digital television can also offer better picture quality, although the difference is more striking when compared with analogue television in the United States than in other countries, which use a different transmission standard.

Digital technology is being used for all three main delivery systems. Satellite delivery was the first, led by the United States. The early success of DirecTV and its rivals was driven partly by factors peculiar to the United States, such as the fury Americans often feel towards cable companies, poor picture quality, and the pent-up demand for multichannel television among people living in uncabled areas. In Europe, Canal+, which started the first national digital satellite service in April 1996, has been successful; the Kirch Group, which launched a digital package in Germany in summer 1996, has not.

Once upgraded to carry digital services, the cable companies will have some advantages over satellite delivery: while each television set needs its own satellite receiver, a single cable point is enough; and some places will never be able to receive satellite signals. But, for older networks, digital requires expensive rewiring.

Over-the-air broadcasting may be the last to switch from an analogue to a digital standard. In early 1997, only Sweden and Britain had firm plans for digital terrestrial television, although in April 1997 the Federal Communications Commission (FCC) set a timetable for launching it in the United States and replacing analogue broadcasting by 2006. Its unique advantage is its portability—a digital set, unlike a cable or satellite one, can be carried around like a pager, a radio, or a portable telephone and thus, one day, can perhaps be combined with one or more of these devices. (It also produces excellent reception: in one British demonstration, the demonstrator pulled out the aerial and pushed a wire coat-hanger into the aerial slot instead. The picture was perfect.) Although digital terrestrial broadcasting will never be able to deliver as many channels as digital cable or satellite, it allows terrestrial broadcasters to offer several channels instead of one. Thus they, too, have a chance to become multichannel broadcasters, no longer tied to the delivery of one "department store" channel, but also able to offer subscription and specialty channels, as cable and satellite operators do.

All the channels available on analogue services will become available digitally as well. Some viewers will switch quickly, but they will need a decoder box or a new digital set. Many viewers—perhaps around one-third—will be reluctant to make the change. That may be because they are elderly or poor or because, in countries with a high-quality analogue standard, the qualitative difference between analogue and digital reception (more akin to that between AM and FM radio than between black-and-white and color television) does not justify the expense.

But, until analogue transmission can be shut down, the switch to digital will not save transmission capacity. To speed the switch from analogue to digital, governments (or existing broadcasters) may, therefore, announce that they will shut down analogue transmission. Mobile telephone networks have already begun to do so: in January 1997, for instance, Britain's Vodafone and Cellnet began to close down their analogue services. First affected were some thirty thousand customers whose telephones dated from before 1986 and who were offered the chance to upgrade at heavily discounted prices. The telephone companies are motivated by the far greater number of calls that can be handled in the same spectrum by a digital as opposed to an analogue system.

The key question will be, who keeps the surplus spectrum? If broadcasters are allowed to sell or re-use the spectrum they save by moving to digital, then, like the telephone companies, they will have a big incentive to throw the switch themselves. But, if spectrum is public property, why should the value go to the companies lucky enough to be using it? This question will lead to plenty of political rows. In the United States, the FCC tried unsuccessfully to argue that broadcasters should pay for the extra spectrum they are being given for digital broadcasting—and then argued that they should finance more public-service broadcasting in exchange for this bonus.

Changes in Content
. . . . .

The revolution now multiplying the number of ways to deliver a television program will bring about another revolution—in content. No longer will the viewer's choice be limited by spectrum, or even by the carrying capacity of analogue cable. In a world where thousands of channels can be delivered to a home, the content of television will clearly be different. But how will it change?

The short answer is that television will become both more varied and more repetitive. First, more experiments will be made, because the barriers to entry into the television business will be lower. Making a 22-minute episode of "Seinfeld," NBC's hit comedy, costs more than $3 million. The big expenses are the four stars, who signed a deal in spring 1997 that will bring them $600,000 per episode, and $1 million in the case of Jerry Seinfeld himself. (NBC was willing to pay because a minute of advertising time on the show sells for an average of $700,000, a clear sign of the continued commercial value of a mass audience.)[11] By contrast, it costs only $3,000 or so to make an hour of programming for Britain's LiveTV, a wacky and tacky cable channel whose star attractions include darts matches played by topless Australian beach girls and a soap opera filmed in its own east London offices in Canary Wharf. In the future, even more eyeballs may watch "Seinfeld," but there will also be many more LiveTVs all over the world.

Second, many more uses for programming that are not thought of today as "television"—for example, video games and internal company broadcasting—will be devised. France's Canal+ already has plans to use its digital package to download software for games. Pro7, a German network, has begun developing in-house channels for businesses to use for pep talks for employees and promotional material for distributors.

Third, viewers will be offered many more chances to see the same small selection of top films and programs. Just as Hollywood now releases movies in a carefully calibrated sequence—first the cinema, then video rental, then video sales, and finally television—so a large proportion of television programming will be released through different "windows," each available at a lower price than the one before, to squeeze out maximum revenue. As a result, the control of exclusive rights to programming will be an ever more valuable competitive weapon in the television business. Content, the raw material of the television business, will be packaged again and again, rather than being broadcast once and then stuck in a library.

Overall, we will see basic changes or development in channel structures; revenue sources; pay-TV, the growth of which will give new importance to rights management; and interactivity experiments.

Changing channel structures

Until the 1980s, most homes—even in the United States—rubbed along with a choice, each evening, of a handful of channels. Like department stores, each of these channels offered a jumble of unrelated products on their schedules: a children's program, say, followed by a sports report, followed by a talk show, followed by the news.

Then along came Ted Turner with his Cable News Network, launched in 1980. This novel idea—nothing but news, around the clock—was greeted with deep skepticism by the rest of the broadcasting industry. But by the mid-1990s, CNN had become an extremely valuable network.

Throughout the 1980s, the spread of cable in the United States, intended to improve local reception, encouraged development of this new sort of channel. ESPN, owned by Disney's ABC, showed nonstop sports events and sports-related news; MTV and Nickelodeon, both owned by Viacom, showed, respectively, music for young people and children's programs. In contrast to the networks' department-store approach, this specialty-store technique operated by creating a market niche and then offering its audiences a continuous diet of its favorite programming. Such markets could be endlessly subdivided: CNN now has seven channels, including one for sports news and one specially designed to show in airports, allowing the costs of programming to be spread over more channels and more viewing time. A newsroom run by a broadcasting network may make three or four one-hour news programs to show every day; a newsroom run by a specialist news business such as CNN can produce programs to be shown around the clock.

Table 3-2   Growth in Television Channels in OECD Europe

	Public	Private	Total	Satellite	Cable channels
1980	44	18	62		
1990	47	84	131	38	
1995	64	180	244	262	463

Source: *1996 Telecommunications Outlook.* ©OECD, 1996, Paris. Reproduced by permission of the OECD.
Note: Television channels with national coverage.
"Satellite" column includes channels distributed by cable networks and ones received from foreign countries.
"Total" column for 1995 includes terrestrial and satellite channels with national coverage, but not cable channels.

In the 1990s, a few new over-the-air channels or networks, such as Rupert Murdoch's Fox network in the United States or Channel Five in Britain, have been launched or planned. But most new television channels have been launched on cable or satellite—and paid for by some sort of subscription. This is true not only in the United States. In Europe too, as Table 3-2 shows, the fastest growth in new channels has been in those delivered by cable and satellite.

Such channels are radically different from the mass-market networks. They readily become brands, for example, with all the opportunities for extension that allows. Not many viewers announce as they switch on the television, "I'm going to watch a CBS show tonight"; but that is exactly what they are saying when they say they are tuning into channels such as CNN and MTV.

This development can be thought of in terms of the rise of boutiques in competition with department stores. Boutiques—strongly branded, with low entry costs, and adept not just at capturing niche markets but actually creating them—can become chains in their own right. The department stores still exist, often by incorporating boutiques, but their dominance over the market has dwindled. Now apply this process to television.

In the United States, where multichannel television—the boutiques—has been available for more than two decades, the share of the prime-time audience taken by the three oldest networks has declined from more than 90 percent in the late 1970s to around 50 percent today. In homes that subscribe to cable or satellite, the change has been much greater: these families now spend two-thirds of their time watching cable and satellite channels, rather than the networks.

In the course of the 1990s, a similar transition has begun in the rest of the world. Between 1990 and 1996, the global number of homes taking multichannel television rose by almost 160 percent, from 90 million to 233 million.[12] Once people have more choice, their viewing habits change. In Latin America, for instance, viewing of terrestrial channels dropped in homes receiving cable, from more than 80 percent at the start of the 1990s to less than 50 percent five years later. Among children, the viewers of the future, the shift to alternative channels has been even more striking.[13] Clearly, loyalty to the networks will continue to erode.

That does not necessarily mean that the networks or national over-the-air channels will disappear—at least, they will not do so quickly. Indeed, the most remarkable thing about U.S. networks has been that they have kept so much of their audiences and advertising revenue.[14] In addition, the pace of change will differ enormously from one country to another. It will be slower in Japan and in parts of Europe, such as Germany or Switzerland, which traditionally have had have good national television, plenty of "free" choice, and more cultural resistance to American imports than is the case in, say, Latin America.

Habit and loyalty will be important. Mass-market television has provided the greatest unifying cultural force since the invention of a common language. Questions such as "Did you watch the game last night?" or "What do you think will happen next on that serial?" form an irreplaceable bond that will help to keep mass-market television alive. Just as a telephone network is more valuable the more people you can reach on it, so a program is more valued the more of your friends want to talk about it the next morning.

By the second decade of the next century, however, people everywhere will have begun to get used to the idea that television means not choice of half a dozen channels but infinite variety. Particular events and programs will still attract immense audiences, but watching television will become a much more individual, personalized experience than it is today.

Changing revenue sources

The new television world of vast choice and fragmented audiences will require different sources of revenue than did the old world of vast audiences and a handful of networks. The result will be an increased share of revenue from subscriptions and from pay-per-view programming.

In its heyday, the old television world was, not coincidentally, marked by mass-market advertising and powerful brands: the age of the brand coincided with the age of mass-market television, an astonishingly powerful tool for promoting brand awareness. So dependent has advertising become on mass-market television that the U.S. networks' advertising revenues have continued to rise even though their audiences have declined.

Advertising, in turn, shaped "department-store" television, creating a clear incentive for the networks to reach as big an audience as possible. Since only a few viewers will buy any given product, advertisers will pay only a tiny sum for each pair of eyes watching the screen.[15] To ensure that advertising revenues cover costs, America's main networks calculate they must reach at least nine out of ten households.

The pursuit of the mass market has been equally important for public broadcasters in other countries, for a different reason. The best political justification for the license fees or state subsidies on which these public broadcasters usually depend is that they cater to all citizens.

The cable and satellite channels, in both the United States and other countries, work on a quite different set of assumptions. Their audiences are minuscule by comparison with the networks. First, they reach fewer homes than does broadcast television—far fewer, in most countries. Second, even in the homes they do reach, ratings for their channels are almost invariably smaller—often far smaller—than those for "free" broadcast television. In the United States, even the best shows on cable networks reach prime-time audiences of only about one to two million viewers, and many reach far fewer people.[16] Even a channel as well regarded as CNN has an average hourly audience in the United States of a mere 400,000.[17]

In Europe, cable and satellite audiences are often smaller still. In Germany, cable and satellite subscribers spend more than 40 percent of their viewing time watching something other than an over-the-air broadcast. But, because Germany has more than forty such channels, each is watched on average for just over four minutes per day per person.[18] As such channels multiply, they tend to cannibalize one another's audiences.[19]

Such fragmented audiences mean that cable and satellite channels rarely reach mass audiences efficiently. But they still can (and do) make money, for two reasons.

First, unlike the networks, they can offer advertisers specialist audiences. ESPN is one of the most efficient ways to reach American men aged between eighteen and fifty-four, while Nickelodeon offers the same advantage in reaching children. Cable also reaches local audiences efficiently, something that advertisers have not really exploited properly, even in the United States.

Second, cable and satellite channels do not rely for revenue only on advertising: they also charge viewers for subscriptions. Two sources of revenue are obviously better than one: in the United States, near the end of 1996, total revenue from subscription television overtook total revenue from broadcast television.[20]

Revenue from subscriptions will probably continue to rise faster than revenue from either advertising or public contributions. In aging societies, subscription revenues bring a particular benefit. Mass-market advertisers want lots of young viewers, who will be big (and adventurous) spenders on consumer goods, and so commercial network channels, at least in the United States, tend to slant their fare to attract that audience. Older viewers, who make up a rising share of audiences everywhere, may find that subscription channels cater more precisely to their tastes.

In the immediate future, subscription revenues will continue to rise faster than those from any other source, not just in the United States but everywhere. That will drive a profound change in the nature of the medium. When payment is made primarily by viewers, rather than by advertisers or through government grants, cable and satellite companies will be able to buy movies and sporting rights and to pay famous presenters. The big money will no longer be with the mass market.

Further ahead, digital technology will transform advertising-financed television. The technical infrastructure that will eventually allow a viewer to order a particular movie or program on demand will also allow the "direct mail" of an advertisement to particular homes, where the viewer will simply press a "buy" button on the remote control to respond. As a result, it will be possible to finance programming watched by a mass audience with personalized advertising, tailored to the tastes and interests of a particular household or viewer.

Pay-TV and the importance of managing rights

In addition to more revenue from subscriptions, television companies will earn more revenue from pay-per-view programming. Viewers will pay for more of their viewing by the program, rather than—as is usual with network and cable channels—buying in bulk. Today, viewers often watch programs they really love at no extra cost, but such bargains will be less common in future. The proposition behind digital television is that programs can be priced to reflect the value viewers place on programs they previously watched for free.

This new pricing arrangement will give greater importance to the management of the rights to the content of shows. Broadcasters have long understood the value of an exclusive right to show a Hollywood movie. But only in the 1990s have live events, and above all live sporting events, become hot properties. Sports rights are, as Rupert Murdoch puts it, the "battering ram" of subscription and pay-per-view television.

Already, experience with digital television in the United States demonstrates that it makes pay-per-view movies easier to offer and more attractive to audiences. Most digital services already offer "near-video-on-demand": a selection of movies starting at frequent intervals. While analogue cable networks in the United States typically offer two or three movies, most DBS services offer ten to twenty starting at more frequent intervals and costing about what people pay to rent movies from a video store. A simple one-button-click ordering system encourages impulse buying. As a result, DBS customers order more than twice as many films per month as do cable subscribers.

For the first few years of digital television, near-video-on-demand will probably be the biggest single use for the new channel capacity. Viewers will be able to watch their favorite television shows more or less whenever they want to: "catch-up television" will rescue all those people who have still not learned to program their videocassette recorders.[21]

One of the most significant consequences of near-video-on-demand will be to increase the revenues (and audiences) for the most popular shows. In several countries (although not in the United States), pay-TV is already close to being the largest single source of movie revenues. Some argue that it will become the largest single revenue source globally by the end of the century.[22] (See Figure 3-1.) This process will be hastened by the development of several different charging bands.

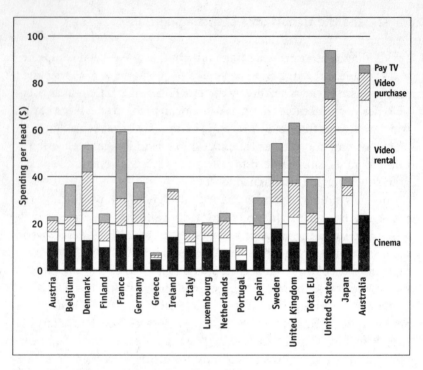

Figure 3-1   Patterns of Movie Spending, 1996
Source: *Screen Digest.*

Many shows will be screened first on pay-per-view, then perhaps on
basic subscription channels, and finally on free-to-air broadcast networks. One consequence will be to increase even more the revenues
that a successful movie—and thus a successful star—can earn.

The pay-per-view premium for live events will be greatest for sport.
The availability of channel capacity will make it possible to show not
just one soccer match or baseball game but all the games that might
interest an audience—and at the time they occur. The technology
makes it easy to charge viewers for watching a particular event or to sell
a "virtual season ticket" to the main matches of the season.

In the United States, DBS offers subscribers a National Football
League Sunday Ticket—a package of all the NFL games played each
Sunday—as well as access to hockey and basketball played in other
parts of the country. In Italy, Telepiu, which launched digital satellite

television in February 1996, developed an "electronic turnstile" that prevents home-town subscribers from watching home games on television. This feature assuages local clubs' worries about losing revenue from supporters at the gate.

Plenty of ideas from sports television may be adopted in other areas: for instance, the concept of a live event as a premium product, so that people pay extra to watch, say, a Broadway show or a key newscast as it happens; the season ticket, which might also be used for opera, theater, or other special series of events; and the multiple cameras, which Kirch, a German broadcaster, uses to allow subscribers watching Formula One motor-racing to switch at will between cameras mounted on the track, in the pits, or on individual cars. But sport will remain special, for three reasons. First, sporting events depend for effect on their "nowness." This quality does not adhere to films and other dramatic entertainment. Miss *The Lion King* on television, and you can always rent the video; miss the Cup Final or the Super Bowl and, given a certain level of testosterone, you will feel a social outcast. Second, sports are among the most popular programming with young adults and especially with affluent young men, a particularly lucrative audience for advertisers. And, third, sporting events do not require expensive product development. The television company does not have to invest in the hit-or-miss business of creating a serial; it merely signs the checks.

The advent of digital television has therefore coincided with a huge escalation in the cost of sporting rights (see Figure 3-2). In the future, television companies will keep a dwindling share of the money they earn from exclusive rights deals. The truly scarce commodity will be not channels of distribution but the sporting events or movies that viewers most want to see. There, the effect of competition will be to bid up costs for a product whose supply cannot be increased: star players or hit movies. Sports people will drive hard bargains, squeezing every penny of profit from the television companies; and movie rights will be sold for ever shorter periods and higher prices.

Interactivity: The content of the future?

Many companies wonder whether television will become the gateway to a new world of interactive services. Of the three possible electronic

· · · · ·

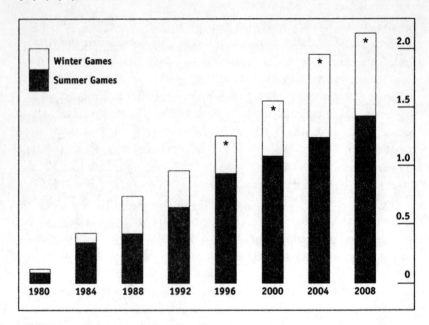

Figure 3-2   Olympic Broadcast Rights
Fees (in billion dollars) as of March 1996.
*Two years earlier
Source: ©*The Economist*, London (July 20, 1996).

gateways from the home—the television set, the telephone, and the personal computer—the television set is by far the most widely owned. Americans own eighty television sets for every one hundred inhabitants, easily the highest penetration in the world.[23] But overall, the world's 5.6 billion people own 1.2 billion television sets—more than one for every five people. That makes the technology more than one and a half times as widespread as the fixed telephone and more than six times as familiar as the personal computer.[24] If sheer familiarity were the only thing that mattered, the television set's claim to be the key to the future of communications would be formidable.

Interactive television has been mooted since the early 1990s, when telephone companies, especially in the United States, feared that cable companies would eat into their markets and began looking to interactive television as a way to make more money from their existing networks. Their idea was that, instead of slouching in front of the set watching a stream of programs at the times chosen by a scheduler, viewers would seize control, using the telephone to download games, to bank money, to book theater tickets, to buy clothes, and so on through any number of other services and goods. The main new product, however, would be "video-on-demand." Viewers would choose a particular television program or movie from an on-screen catalogue; they would then watch their selection at whatever time they wished and would be able to fast-forward, reverse, or pause, just as with a videocassette. Later, they might be tempted to pay for a second viewing.

Many trials were set up to see how such a network might work, some by telephone companies and some by cable-television companies. In Orlando, Florida, Time Warner hooked up four thousand families to a service offering home shopping, movies on demand, and video games. In Britain, BT connected twenty-five hundred families in Colchester and Ipswich to a similar service. By mid-1994, interactive trials were being conducted by a variety of cable companies, telephone companies, power companies, and twenty-three program providers (such as MTV and the Sega Channel)—in the United States alone.[25]

The trials (described in an article in *Wired* magazine memorably entitled "People Are Supposed to Pay for This Stuff?") found that the sums simply did not add up. Instead, the near-video-on-demand services offered by ordinary digital satellite broadcasters have proven to be a rough-and-ready way of persuading viewers to spend much more on movies, without requiring the enormous investment required by truly interactive television. Interactive television simply requires too much up-front investment in fiber-optic links reaching most of the way to the home. In the words of John Malone, chief executive of Tele-Communications Inc., a big American cable company, "In the short run the economics of providing a consumer with a device capable of interactive video and display on the TV set is just too high to expect any kind of reasonable market penetration at the prices that would have to be charged."[26]

So interactive television, in the image the telephone companies held of it in the mid-1990s, is dead until the costs fall much further. But the large capacity of digital delivery will allow broadcasters to develop other services that will include elements of interactivity. Children, for example, will watch adventure stories that allow them to change the plot by switching channels. Cable services already offer computer games that allow groups of subscribers on different parts of the network to play against one another. In addition, the transactional aspects of interactive television will survive. In Britain, BT joined forces with BSkyB in May 1997 to develop ways to offer shopping, banking, and other transactions on a television set.

Interactivity will also be enhanced by the growth of the Internet and the plans for digital terrestrial television broadcasting. Many people now have a strong interest in proving that, in a digital world, the convergence of the television and the Internet has commercial value.

Television networks, for example, hope their Web sites can prevent the Internet from siphoning off their audiences. NBC was astonished when, in summer 1996, its Olympics Web site got 7.3 million hits in one day. Sportzone, a site run by ESPN, has become the leading Internet site for sports news—and for advertising for sports equipment. John Birt, head of Britain's BBC, talks of the Internet as the "third medium," after radio and television.

Manufacturers of PCs and computer software have also acquired a strong interest in the possibilities of convergence. The three giants that dominate the computer business in the United States, Microsoft, Compaq, and Intel, have benefited greatly from the boom in PCs. Americans spent $19 billion on computers in 1996—compared with $10 billion on television sets. They are keen to benefit as television becomes a more versatile medium. They see that the biggest potential for bringing together television and the Internet will depend on the way digital television develops. They have been lobbying the television industry to ensure that digital broadcasting in the United States takes a form that allows PCs and TV, and the material on their screens, to be easily interchangeable.

In addition, Microsoft has teamed up with NBC to launch MSNBC, an on-line and cable service offering six television channels that can be watched on a PC as part of MSN, the Microsoft Network. At the start of 1997, the service had 1.8 million subscribers and was beginning to offer live coverage of business news.

More important, Microsoft acquired WebTV Networks in April 1997. This company had developed a way for people to use a television set as a simple PC, sending and receiving electronic mail and exploring the World Wide Web. The WebTV was priced at around $300, or well below the $1,000-and-up price tag of a new computer. Sony sold 35,000 models in the first two months after launch, a figure that CD players took a year to reach.[27] If Microsoft is right, the Web TV may turn out to be a prototype for bringing the Internet to the television screen.

Will all this effort lead to something that viewers actually want? Perhaps: thematic channels and Internet access could be used to create something like an interactive magazine program for the innumerable niche markets that printed magazines now reach. A dogs channel, for example, could devote different nights to different breeds: a bulldog night, a retriever night, a mongrel night. Or a group of universities could acquire a channel and each take one evening a month to distribute news to alumni and raise money.[28]

With appropriate set-top boxes, dog-channel viewers could summon information on dog training or breed championships or use a digital camera to hold video conferences with other dog-loving viewers and their mutts. Included in the dog-owners' subscription might be access to a screen-based bulletin board carrying specialized advertising, much of it classified. The set-top boxes would also feed television companies precise information about who was watching what, allowing advertisers to target the right viewers with the right dog-food and other promotions.

But the main use of such interactivity will almost certainly be in offices, where people may use their screens to watch breaking news, take part in distant training sessions, or monitor conferences without traveling. In the long run these uses—pulling together aspects of television and of the Internet—will probably be among the fastest growing uses of the television.

Changes to the Industry
· · · · ·

Just as increased capacity of the telephone network is altering the structure of that industry, so much the same is happening in television. Three trends are striking, all of them echoing trends in telecommunications: first, a shift away from public-sector monopoly and regulation to a competitive, privately owned industry, more like any other; second, new alliances among media and content providers, some of them cross-border, others aiming at greater vertical integration; and third, new "gateways"—points of scarcity or bottlenecks

. . . . .

where one company can gain a competitive advantage over others—reflecting a new balance of power.

The future of public broadcasting

Aside from the United States and a few other countries, the world's biggest broadcasters have long been public corporations. Some, such as Britain's BBC and Japan's NHK, have ownership structures that carefully distance them from government control. Others, especially in developing countries, do not. All depend mainly on public money, license fees, or levies in some form or other.

Now, many of these giants have suffered the same hemorrhage of viewers as have America's networks, which are their nearest equivalents in terms of audience and attitude. For example, Germany's two public networks, ARD and ZDF, together lost 44 percent of their audience between 1990 and 1994; Spain's TVE lost almost 20 percent of viewers in the same period; and NHK, Japan's public broadcaster, found in a 1994 survey that, although 90 percent of people in their fifties and sixties watch its programs at least once a week, the "contact rate" dwindled with each succeeding age group to a mere 50 percent among people in their twenties.

In time—although perhaps not until the end of the first decade of the next century—the nature of public broadcasting will have to change. As channels multiply, governments will find it increasingly odd to subsidize one or two of them. In addition, it will be possible to measure quite precisely who watches what, which may lead people who do not watch public broadcasting to demand to opt out of paying for it.

Countries wedded to the idea of public subsidies for television may instead underwrite particular programs considered nationally desirable, rather than financing entire channels. Public television will thus become part of the general state budget for subsidizing the arts, rather than a separate institution.

New alliances

The *Wall Street Journal* publishes an annual chart of the links among the world's largest entertainment groups. The result looks more like a

plate of spaghetti with every passing year. No industry is more inces-
tuously intertwined than the entertainment business—especially the
television industry.

To take one example: TCI, America's second-largest cable company, which was nearly taken over by Bell Atlantic, a regional telephone company, and which owns Liberty Media, which claims to be
the world's largest programmer, owns part of Time Warner, America's largest cable company, which in turn now owns Turner Broadcasting, America's third-largest cable company, and which has had
an acrimonious joint venture with USWest, another regional Bell.
Simple, really.

Such cat's cradles are not unique to the United States. For example,
France's pay-television broadcaster, Canal+, has a stake in Vox, a German television channel, and is part-owned by Havas, a French media
group that also has a stake in CLT, a Luxembourg broadcaster, whose
television interests were merged in 1996 with Bertelsmann, Germany's biggest media group, which has a stake in Premiere, a German
pay channel in which Canal+ also had a stake (sold in July 1997).

These alliances reflect three characteristics of the television industry.
First, the industry remains uncertain about which technologies will be
winners. Thus TCI part-owns PrimeStar, a satellite broadcasting venture, and MCI has a stake in News Corp. Second, television companies
ally themselves with other companies within the industry in response
to the difficulties of breaking into foreign markets, difficulties that
stem in part from regulations aimed at discouraging foreign companies from gaining control of a country's main media industry, but also
from the strong cultural aspect of television. Thus, to launch digital
satellite broadcasting in Latin America, News Corp. teamed up with
Brazil's Globo, Latin America's biggest entertainment empire, and
with Mexico's Televisa, the biggest television company in the Spanish-
speaking world.

Third, the owners of distribution systems seek to own content and
vice versa. In the complex evolution of the industry, content owners
have the ultimate scarce resource, but they can still find themselves in
a weak bargaining position if they do not have guaranteed access to
delivery channels. Thus Rupert Murdoch waged a high-profile war with
Time Warner in 1996 and 1997 to get his Fox News channel onto the
main New York cable network.

New gateways

In a world with thousands of channels, there are still opportunities to exert market power—but different ones. They include the set-top box and encryption technology; the electronic program guide; and access to desirable content.

The set-top box

In the United States, where digital satellite broadcasting has been launched by companies outside the television mainstream, little political debate has taken place over the "conditional access" system—the combination of set-top box and encryption technology that allows a broadcaster to make sure that only those who subscribe to a service can watch it. In Britain, BSkyB plans to launch the first digital service in 1998. Other broadcasters fear that BSkyB may thus control their access to viewers. They argue that, once viewers buy a set-top box designed to watch BSkyB, they will be unlikely to acquire a second to receive programs from a rival. Technically, one box can be used to decode the signals from many companies' channels. But the broadcasters who are first into the field have a clear commercial interest in using proprietary software and forcing future competitors to pay them for distribution rights. A similar debate has gone on for years in the American cable industry over the terms on which cable companies carry programming produced by rivals.

The electronic program guide

"Just as software and system services drive the computer industry," says Nicholas Negroponte, an academic at MIT's Media Lab, "programming and intelligent browsing aids will drive the television industry."[29] Adrift in a sea of channels, how will viewers find what they want to watch? The answer will be search engines, rather like those used on the Internet.

Broadcasters will fight to make sure that their own particular offerings are listed at the top of any electronic program guide. In this, they will be stealing an idea from the airlines, which discovered long ago that the companies listed on the first page of a computer reservation system got the most bookings, even if services on later pages met customers' requirements better.

Access to content

The most intractable monopolies are the unique events—mainly sports—valued far more by viewers as live television than as tapes. Given that many countries think of sporting events as part of their national heritage, the control of sporting rights by pay-television companies will lead to political opposition. Viewers do not like being forced to pay for something that has previously appeared to be free; and mass-market television channels, faced with increasing competition, will make a fuss.

In the United States, the networks carry most big sporting events, such as the Super Bowl and the Olympics. But there have still been rows: in August 1996, for instance, the National Basketball Association sued America Online to stop it from carrying news of games while they were taking place. The previous month, the NBA had persuaded a federal judge to close down a sports-updating service offered by Motorola to its pager customers.[30]

In Britain, the arguments have ended up in Parliament, where in March 1996, the government legislated to prevent the rights to eight popular sporting events from being sold exclusively to subscription television (which, in practice, meant to BSkyB). In Spain, in early 1997, the government had plans for legislation to prevent pay-TV companies from obtaining exclusive rights to football matches.

Most abuses of competition can be checked by appropriate regulation. Sporting events may turn out to be different. For the ultimate monopolists are not the media companies, but the teams. Television companies can play off seven or eight Hollywood studios against one another. But most countries (apart from the United States) have only one national sporting passion—football—and only one national league. In the long run the players and the clubs hold all the cards. The television companies are more likely to be their servants than their masters.

Global Television

· · · · ·

The question raised by the *New York Times* all those years ago remains important. Time will be the biggest constraint on future demand for television at home. So, if the television business is to grow, television companies will not only have to find ways to charge existing viewers

· · · · ·

**Table 3-3   Average Television Viewing Time
per Household per Day, 1995**

Country	Hours
Australia	3.00
Canada	3.30
France	2.91*
Japan	3.88
South Korea	2.52
United Kingdom	3.50
United States	4.00

*1994
Source: *1996 Telecommunications Outlook.* ©OECD, 1996, Paris. Reproduced by permission of the OECD.

and new domestic markets for television, they will have to attract new audiences in parts of the world with lower viewing rates and a lower quality of locally made programming. (See Table 3-3.)

Access to television, so widespread in the rich world, is growing at astonishing speed in many developing countries; and, as transmission hours and program choice increase, so do viewing hours. In many parts of Asia, access to television has been rising at around 20 percent per year since the early 1980s, bringing many of the revolutions in marketing and the broadening of horizons that took place in the rich world in the 1950s and 1960s. Advertising stimulates a demand for branded goods and encourages consumers to try new products, for example; the spread of television through developing countries is, therefore, likely to prove a powerful force for economic change.

For the moment, American television companies are expanding vigorously abroad, racing to develop services in the booming markets of Asia and Latin America. U.S. companies have an advantage over competitors in other countries partly because American culture seems to travel better than any other; in addition, America's television companies have lived longest in a world of many channels and lively competition.

All the big American television companies sell their channels abroad, and some build their own distribution networks. For example, TCI is building a network to carry cable television and telephone calls with Sumitomo in Japan. Most dramatic of all has been the push by Rupert Murdoch's News Corp. to create a global satellite-television empire. The company is involved in joint ventures in Europe, Japan, and Latin America and owns Star, a satellite-television service based in Hong Kong.

Building a global television business is not quite like building a global oil empire, as is clear from Mr. Murdoch's failure to succeed with a single pan-Asian service. National television markets are different—people follow different sports, laugh at different jokes, are shocked by different things, care about different news events. Some television programs seem to travel: "Baywatch" is a success in India (dubbed in Hindi); and Britain's thirteen- to sixteen-year-olds rate "Friends" their favorite program, marginally ahead of "Neighbours," an Australian soap. But most television material needs adapting to fit local tastes. Several American television companies, such as Disney's ABC, have built their own production facilities in Asia to make programs tailored to local markets. An example: a new "Sesame Street" has been specially developed in Israel with Israeli and Palestinian Muppets and other characters.

Most global television will always be paradoxically national. Indeed, a multiplicity of channels will allow television to cater more precisely to regional and local tastes. More channels will reach more viewers: for example, JSTV, a Japanese satellite channel, has twelve thousand subscribers among Japanese business families living in Europe. More services will be provided, as on ARD, Germany's main public broadcaster, which transmits its entire evening news program across the Internet with full sound and video, aiming it partly at Germans living abroad, and especially in the United States. The world's shifting populations will thus be able to keep in touch with their cultural roots.

Everywhere, more television from other countries will be available. Where local television is mediocre, people will watch imports from abroad, mainly from the United States. Competition, and the lower cost of making basic programs such as local news, will vastly improve the quality of television, especially in developing countries. Imports will also have cultural effects: they will influence expectations, not just of living standards, but of political and social freedoms. Programs that help to glamorize Western lifestyles, especially those that display desirable goods, will raise economic aspirations. In countries such as India and China, television may thus help to drive the market economy. In China, it may even create a taste for democracy.

In the rich world, the dominant influence on television will be the Internet. Apart from competing for viewers' time, the Internet will change some of the things that television can do. One of the most extraordinary features of the Internet, the mass of experiments it fosters, will affect television, as well: as the next chapter explains, one of the key uses of the Internet is as a broadcast medium.

In a decade's time, television will have become a far more diverse medium and will be watched in many more places than it is today. Moving video pictures, professionally assembled, will become the only common theme (and some of the assembly will be amateur). Television, of a sort, will be delivered to handheld gadgets (perhaps one that gives drivers stuck in traffic jams a glimpse of road conditions ahead). Television will merge with computer games; it will deliver news programs on specific issues on demand to a desktop screen; and it will develop an entire subsidiary industry of in-house television for companies, broadcasting to their staff, their suppliers, their distributors, and their customers.

But there will also continue to be television that recognizably descends from what people watch today. This familiar programming will be delivered, not to a computer, but to a distinct sort of machine used for particular activities, most of them fairly passive and most linked to entertainment rather than to shopping or an information search. Indeed, in the first decade of the next century, the quality of sets will improve, as wide screens become commonplace, bringing big-screen quality—and, incidentally, making the television set even more different from a PC than it is now. Digital delivery and a flat screen that hangs on the wall will make television feel more like genuine home movies.

Coming home in the evening, the average person will still sprawl on the couch and watch (though some earnest souls may surf the thousands of channels while they ride their exercise bikes at the gym). Viewing choices will increase; some program costs will rise; picture quality will vastly improve; and people will choose from electronic menus. But, well into the first quarter of the next century, the great leisure activity of the twentieth century will still be essentially just that, a leisure pursuit: passive viewers we will remain.

4

The Internet

In January 1997, a few weeks before President Clinton delivered the first ever State of the Union address to be carried live by Internet radio, French President Jacques Chirac opened his country's new national library. As he toured the building, the president was shown a computer "mouse." Astonished by this novelty, the president gazed at it in wonder.

The technological gap between the United States and the rest of the world yawns nowhere more widely than in computing and communications, and especially in the use of the Internet. More than half the world's Internet users are in America. And yet, the growth of the Internet has been the most astonishing technological phenomenon of the late twentieth century. In 1990, only a few academics had heard of it. By 1997, on some estimates, fifty-seven million people were using it; if those who use it only for electronic mail are included, the total jumps to seventy-one million (see Figure 4-1).[1] For a decade, from the mid-1980s to the mid-1990s, the number of people tapping into the Internet doubled every twelve months. By early in the next century, the number of users may pass 300 million, catching up rapidly with the world's current stock of 700 million telephones.

The Internet is remarkable not just for its popularity but for its innovative force. Invented neither by the telephone nor by the television industry, it was, instead, born in the world of academic research, and its spread has been powered by spontaneous demand from millions of users, rather than by corporate marketing departments. It has unleashed an astonishing torrent of new ideas, not just in the hardware and software that people need to use the Net, but also in services that companies sell over it (discussed in the next chapter), and it has created fortunes, such as Netscape Communications, founded in 1994 and floated the following

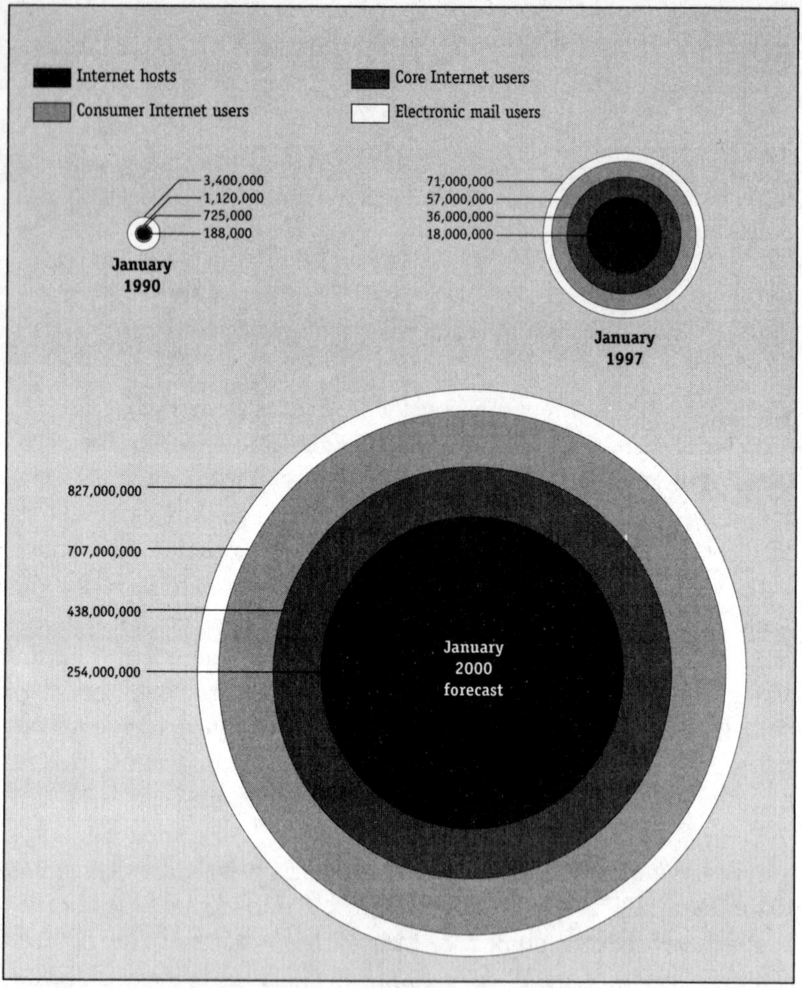

Figure 4-1   How Many People Use the Internet

Note: Core Internet: users of computers that can distribute information through services such as the World Wide Web.

Consumer Internet: users of computers that allow access to information through services such as the World Wide Web.

Source: Data from Matrix Information and Directory Services, Austin, Texas http://www.mids.org.

year with a market capitalization of $2 billion. Many of its innovations may lead nowhere, but some will transform communications. Anybody can now see that the Internet will be important, even though nobody can be sure what it will be like twenty years from now.

The Internet is a product of the same technological revolutions that have slashed the cost of delivering telephone calls and television programs and multiplied the capacity of both sorts of network. Unlike the telephone and the television, however, the Internet has no established principal use. Instead, it has many uses, including carrying telephone calls and television programs. An open conduit, it is capable of transmitting anything that can be put in digital form.

Thus the Internet is not dominated by a single large industry, forced to defend a pre-revolutionary cost structure or market. It therefore offers a glimpse of the communications future: a world where transmitting information costs almost nothing, where distance is irrelevant, and where any amount of content is instantly accessible.

But it affords only a glimpse. The Internet is a merely a prototype of something much more sophisticated. As uses for the Internet multiply, its shortcomings become more noticeable and limiting. By the start of 1997 came the first hint that the Internet's astronomical growth might be settling down to something merely astounding: instead of more than doubling every year, the number of computers connected to the Internet rose by a "mere" 70 percent in the year ended January 1997.

This slowing of growth may not be significant—exponential growth is always unsustainable. The challenges to the expansion of the Internet, however, will be more apparent by the turn of the century, when answers should be available to two key questions: first, will the Internet remain something used mainly in business, or will it, like the telephone, become part of everyday home life? And, second, will it be able to provide a high-quality network—fast, reliable, secure—for high-value services?

For the Internet to rival the telephone and the television as a tool of mass global communication, it will have to be accessible in the living room as well as on the desktop; reliable, secure, and easy to use; truly multinational, rather than mainly American; and available to rich and poor, around the world.

The next few years will show the extent to which the Internet can acquire these qualities without losing its original exuberance. For the Internet to emerge as the information superhighway, its main services must be as accessible to a grandmother as to a geek. It must combine the quality of service of the telephone with the fun of television, at prices that rival both. Above all, until people cease to be aware of it as something special and complicated, it will not have found its future.

How the Internet Took Shape

. . . . .

Many people think of the Internet as a child of the 1990s. In fact, its roots and many of the ideas behind it go back to the late 1960s. Even the word "hypertext" goes back to the early 1960s, when it was coined by a dreamy utopian called Ted Nelson who came up with the idea while studying sociology at Harvard. Well before the invention of the World Wide Web in 1989—which really made the Internet accessible —many of the Internet's most important characteristics had already developed.

Somewhere between 16.1 million and 18 million "hosts"—the main computers that store information—were connected to the Internet at the end of January 1997 (see Figure 4-2).[2] This does not indicate the number of people using the Internet: each host could account for between one and several thousand people. But these enormous numbers are relatively recent. Only since 1993 to 1994, with the development of the Web, has the Internet acquired a large market. Only since then, too, have most companies begun to look at the Internet as a marketplace. And only in the final years of the century are some of the capabilities that may shape the Internet's future beginning to come into use.

The origins of the Internet

A glance at the Internet's history will help to isolate its defining characteristics. More important, it shows just how different the Internet is from what went before and why it has such extraordinary versatility. Indeed, more perhaps than with any other new technology, the Internet's history is also the key to its future.[3]

The Internet, more or less from the beginning, has exhibited a unique combination of features: it is a product of the public sector; its concept was dismissed by the telephone industry; it was built on a single stan-

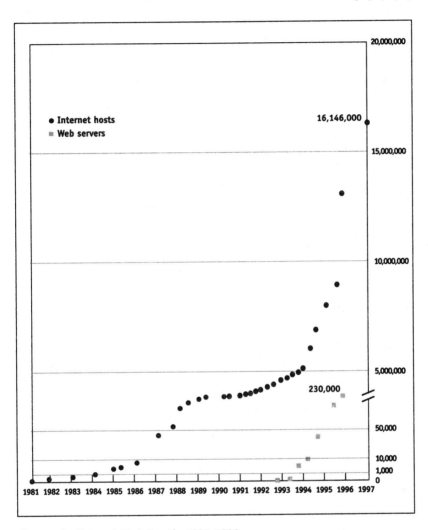

Figure 4-2   Internet Host Growth, 1981–1996
Source: Adapted from TeleGeography; additional data from Network Wizards
(http://www.nw.com) and Matthew Gray.

dard; it is and always has been public property; it has no central command; and it is predominantly American. Finally, although the World
Wide Web sprang fully into being only in the early 1990s, it has been the
most important single force for combining the Internet's idiosyncratic
virtues into a new and powerful medium of mass communication.

. . . . .

Product of the public sector

The Internet is a product, not of entrepreneurship, but of academic research, publicly financed. In the 1960s, computers were large, expensive, and scarce. The American Defense Department's Advanced Research Projects Agency (ARPA) backed an experiment at a small company in Cambridge, Massachusetts, called Bolt Beranek and Newman (BBN), to connect computers across the country as a way of multiplying their power. By building electronic links between machines, researchers in different universities could share resources and results. In 1969, the computerized switch that was the key to the prototype network was installed at UCLA. It was the basis for a nationwide network, called ARPANET, which initially linked four university sites and existed until 1990.

The Internet, which grew from ARPANET, is now a global network of commercial and noncommercial computer networks. In other words, it is a network of networks, some of them immense, others small. Linking them is a spider's web of leased telephone lines and, increasingly, capacity operated directly by the world's big telephone companies. But it was only in 1994 that the number of commercial computers connected to the Internet overtook the number of computers at academic organizations.

Ignored by the telephone industry

The Internet demonstrates a general principle common to big inventions: they never come from the industry that they supersede. The railway industry did not invent the automobile. The telephone industry did not invent the Internet. Indeed, it turned it down.

Two key aspects of the Internet are that it is distributed and that it is packet-switched. Both these revolutionary ideas sprang from work by Paul Baran at the Rand Corporation in the early 1960s. In the tense years of the Cold War between the Soviet Union and the West, Baran was interested in designing a communications system that could survive a nuclear attack. Vulnerable communications were a threat to peace, he argued, because if communications could be easily disabled before retaliation to an attack could be launched, pre-emptive strikes became more likely. Baran came up with the idea of distributed networks, designed like a fish net, rather than the centralized network then typical of the telephone system. If one link in such a network were to be knocked out, a message could always travel by an alternative route.

To make that easier, Baran came up with a second idea: split the message into fragments and send each one separately. Computers, unlike telephones and television, had always been digital, handling information in a stream of digital ones and zeroes. Baran envisaged a network of unmanned switches, or computers, reading the address on each message and sending it on—"hot-potato routing," as he called it—to the next junction on the network. At the next junction, the process would be repeated, until the message reached its destination. There, other information encoded in the message would allow it to be reassembled into its original form.

Baran took his radical ideas to AT&T. The company was uninterested. But Baran's ideas subsequently became the core of Internet technology. One enormous advantage of what has come to be known as "packet-switching" (the phrase, unlike most Internet phraseology, was coined not by an American but by a Briton) was its frugality. A telephone call is "circuit-switched"—sent across a circuit that, for the duration of the call, is established and made available, end to end, for that call alone. That is wasteful, since gaps will occur in the conversation when no signal is transmitted. But it ensures a given amount of transmission capacity, or bandwidth, that not only allows both callers to talk and listen at the same time but also provides enough to spare in order to guarantee the quality of the connection. By contrast, the fragments of an Internet message do not need to take up an entire circuit: each can be slotted into a continuous stream of ones and zeroes, allowing a far more efficient use of transmission capacity.

Now packet-switching has become the main way to send data around the world. And traffic traveling over the Internet, the dominant data network, is growing roughly ten times faster than international voice telephone traffic and, by the end of the century, may overtake it on some routes.

A single standard

The Internet is built on a single standard because the U.S. Defense Department, the world's largest buyer of computers in the late 1960s, was compelled by law to be fair to competing computer manufacturers.

. . . . .

Different machines proliferated, each with different operating standards. A fundamental requirement for connecting computers was to overcome these incompatibilities. The first real standard (or "protocol") was not completed until 1971. And TCP/IP (Transmission Control Protocol/Internet Protocol), which now lays down the format in which all the data sent over the Internet is packaged, was designed mainly by Vinton Cerf and Jon Postel in 1974 and formally introduced in 1983, the date usually taken as the Internet's proper starting point.

To talk to one another, computers need not just a physical link but also a common language. The common protocol, in effect, links any number of computer networks so that they can behave as a single network. From a PC to an Apple Mac to a mainframe to any other computer, across the Internet all can converse in one language.

Public property

The Internet's standard is public property, available for anybody to use for free. In that sense, it resembles the English language more than, say, Microsoft's near-ubiquitous Windows operating system, whose private owner makes money (lots of it) from its popularity. Because it grew out of the worlds of public funding and academic research, the Internet protocol is available for anybody to use, without a license, payment, or permission. The initial financing came from ARPA, making this arguably the American military's most influential peacetime project. But throughout the early 1980s, the main users were the large research universities. They strung together their networks of computers (using Ethernet, a local area network that had been designed by Bob Metcalfe back in the early 1970s), and then hooked them into the Internet.

This wonderful open quality has stimulated a great surge of creativity. While the telephone network is controlled by the telephone companies, which set rules about what can be connected to it and on what terms, the Internet puts power in the hands of the user.

No central command

The Internet has no central command. Its traffic runs mainly over lines leased from telecommunications companies, but they neither manage nor take responsibility for it. Indeed, nobody owns the Internet, runs it, maintains it, or acts as gatekeeper or regulator. Anybody can send a message across it or create a "site," a file of information, held on a computer, to which other users have access.

The few decisions that are taken centrally—on issues such as refining the protocol or establishing principles for allocating Internet addresses or "domain names"—are taken by a handful of mostly American engineers and scientists who run, sometimes part-time, the Internet Society and a few other bodies such as the Internet Assignment Numbers Authority and the Internet Engineering Task Force. But they act as self-appointed guardians, rather than owners, and their approach is determinedly informal and permissive.

American dominance

The Internet's early development took place almost entirely in the United States, and it remains predominantly American.[4] This unmistakable American dominance is shrinking, however: in early 1997, just over half of all hosts—and so, possibly, of all users—were in the United States; a year earlier, the proportion had been nearer three-quarters.[5]

The dominance, not surprisingly, affects the language of the Internet. An analysis of hosts in 1996 found that English was more than twice as common as German, the next most commonly used language.[6] In early 1997, of the dozen countries with the highest number of Internet hosts per thousand of population, four were English-speaking and a further five were countries of northern Europe and Scandinavia where English is widely used as a second language (see Figure 4-3).[7] French President Chirac spoke accurately when, on the occasion of his first confrontation with a "mouse," he dismissed the Internet as an "Anglo-Saxon network."

The Internet will become less Anglo-Saxon as the number of sites in other countries grows. But the culture and not just the language of the Internet is also strikingly American. Its quirky blend of technocratic individualism, egalitarianism, and passionate resistance to government control all seem to many foreigners quintessentially American. It is a powerful tool for exporting American ideas and ways of doing things to the rest of the world.

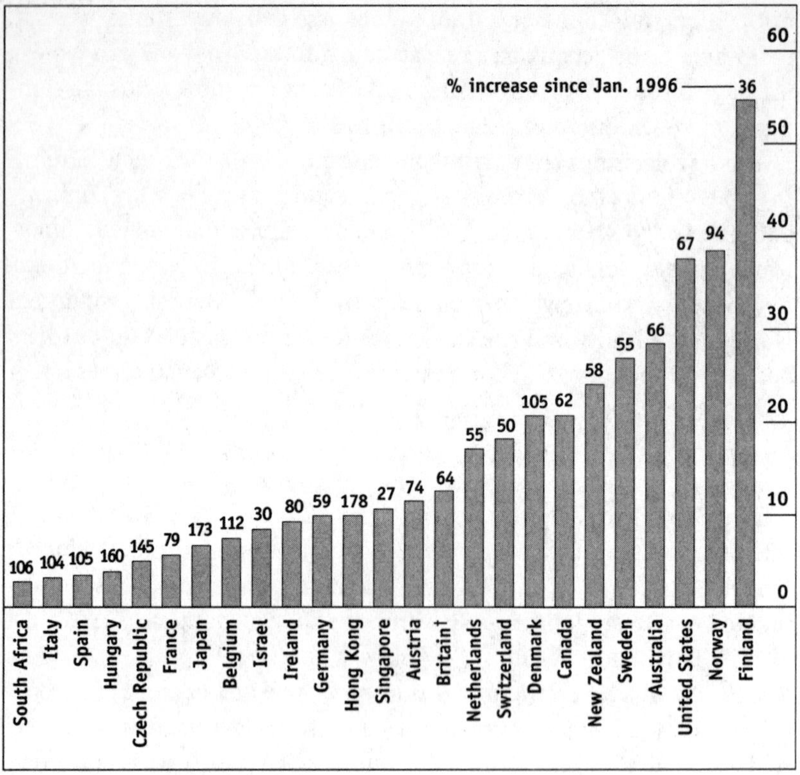

% increase since Jan. 1996 ——— 36

60
50
40
30
20
10
0

106 104 105 160 145 79 173 112 30 80 59 178 27 74 64 55 50 105 62 58 55 66 67 94

South Africa · Italy · Spain · Hungary · Czech Republic · France · Japan · Belgium · Israel · Ireland · Germany · Hong Kong · Singapore · Austria · Britain† · Netherlands · Switzerland · Denmark · Canada · New Zealand · Sweden · Australia · United States · Norway · Finland

Figure 4-3 Top Internet Countries
Hosts per population of 1,000, January 1997.*
*Includes all ending ".com", ".net", and ".org", which exaggerates the numbers.
†*The Economist* estimate.
Source: ©*The Economist*, London (February 15, 1997).

The World Wide Web

What finally transformed the Net from an academic billboard into a popular craze was the World Wide Web. Invented in 1989 by Tim Berners-Lee, a British researcher at CERN's European Laboratory for Particle Physics in Switzerland, the Web only sprang to life four years later when Marc Andreessen, a twenty-three-year-old programmer, and his colleagues at the University of Illinois came up with Mosaic, the multimedia Web browser.

The Web has allowed the Internet to work in new ways. First, it has brought multimedia to the Internet. The Web allows the display of colorful pictures, music, and moving images as well as data and text. The result has been to make the Internet much more accessible and fun to look at.

Second, it introduced hypertext. Essentially a tool for cross-reference, hypertext allows users to move directly from a word or phrase highlighted on the screen to related information that may be stored on a different computer in another part of the world. Groups of related documents can thus be used together, even if they are stored in quite different places.

Third, browsers made possible a quantum leap in the simplicity of both hypertext and the Web. Marc Andreessen's Mosaic software, now commercialized as the Netscape Navigator, allows users simply to point a mouse at a hypertext word or a picture and to click on it, thus opening up a new file. With the creation of Mosaic, the Internet completed its migration from a computer scientist's research tool to something that a toddler can use.

By January 1997—fewer than four years after the birth of Mosaic—more than a quarter of a million Web sites were available for browsing. So popular has the Web become that many people now talk of it as synonymous with the Internet itself. In fact, a growing number of Internet applications are separate from the Web, such as electronic mail (the biggest use of all), telephony, and the various new ways to publish or "push" information to users, rather than waiting for them to come and "pull" it. But the Web made the Internet user-friendly. And this new accessibility came at a time when a number of other forces were driving forward use of the Internet.

The drivers of the Internet

With the coming of the Web and the Netscape browser, the use of the Internet began to change. Between 1993 and 1994, the Internet seized the public imagination. In 1995, it was still possible to write a book on the future of the computer and communications industries and barely mention the Internet: Bill Gates, boss of Microsoft, devoted about twenty pages to the topic in the first edition of *The Road Ahead*. But, by fall 1996, nearly twenty million Americans were logging on to the Internet weekly, four times as many as a year earlier.[8] In the United States (but in few other countries), the Internet has become an accepted part of many people's working lives.

What brought about this transformation? Among the driving forces have been the Internet's low cost; the spread of the PC and of local-area networks in companies; the Internet's open standard; and the development of on-line services.

The cost of using the Internet

The Internet demonstrates a world in which distance does not affect the price of communicating, nor, in some places, does call length. Beyond a doubt, the Internet owes much of its popularity to the fact that it is such an inexpensive way to communicate.

Why is it so much less expensive than the telephone? In part, because the Internet's packet-switched technology crams more messages into the same transmission space than does a standard telephone call. Communicating anything via the Internet over however far a distance is a bargain, compared with using the telephone itself.

But that is not the whole story. To understand how the economics work, look at the ways people gain access to the Internet. They have two main options. First, they can use a line leased from a telephone company, as do most offices that are linked to the Internet. Because the office (or its Internet Service Provider, or ISP) leases a fixed amount of capacity, the charges remain the same, however much the capacity is used or wherever the call goes. Second, users can dial up the Internet from a modem, as do most people calling from their homes. People dial their local ISP, paying the cost of a local telephone call to do so. After the initial call, their Internet connection is carried on a leased line, for which they pay the ISP a subscription fee.

So the cost to subscribers for using the Internet is affected by two things: the cost of leased lines, which influences what service providers charge, and the price of local telephone calls. As Figure 4-4 shows, these can vary widely between countries. The United States has the world's largest and most cutthroat market for leased telephone lines, which has held down their cost. By contrast, ISPs in some parts of Europe and in Japan tend to face high charges, limited capacity, and an obstructive telephone monopoly. One study found that the average price for leased-line Internet access was 44 percent lower in countries with competitive telephone businesses than in those with a national monopoly.[9] International private circuits in

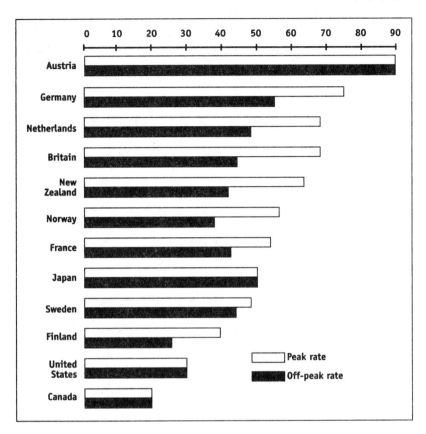

Figure 4-4   Net Costs
Cost of Internet use in U.S. dollars per twenty hours on-line per month (including both telephone and Internet access charges), August 1996.
Source: OECD, ©*The Economist*, London (May 4, 1997).

Europe cost roughly ten times as much as American domestic circuits of the same length.[10]

In most countries, local calls are charged by duration—the longer you talk, the more you pay. But in much of the United States and in a few other countries, local calls are either "free" (the price is included in the basic subscription) or cost a flat amount per call. That makes a huge difference to the cost of using the Internet. In countries where calls are charged on the basis of duration, users can pay up to ten times more for a few hours of access than do users in countries with flat-rate or free local calls.[11] A German using the Internet for twenty peak-time hours a month would pay nearly four times as much as a Canadian.[12] Not surprisingly, Internet use is highest in countries in which people pay the lowest charges for lengthy local calls and vice versa.

"Free" Internet use brings problems, as America Online discovered in the first months of 1997. Its subscribers began to treat its service rather like television—switch on and stay on. Congestion and delays resulted. The worse the congestion, the greater the temptation for people who had already switched on to stay on the line (to keep a connection), thus making congestion even worse.

But expensive Internet use is equally undesirable. Governments often talk about the need to invest big sums in building an "information superhighway." Tony Blair, Britain's Labour leader and Prime Minister, talked during the 1997 election campaign of giving every British school child his or her own Internet address. Sometimes politicians want the money to come from taxpayers; sometimes from the telephone companies. The Internet provides a perfectly good model of a superhighway, and the best thing governments could do to foster its use would be to encourage competition among the companies that provide its infrastructure, in particular the telecommunications giants. Such competition will have far more effect than any amount of highway-building—and will be more efficient, too.

The spread of the PC

Bill Gates has described the popularity of the Internet as "the most important single development in the world of computing since the IBM PC was introduced in 1981."[13] The Internet has certainly vastly increased the value of the PC, but the relationship also works the other way around: the widespread adoption of the PC prepared the ground for the Internet. It taught people to think differently about computers and gave them the tools they would eventually need to use the Internet.

In the 1970s and early 1980s, desktops carried "dumb" terminals that could do nothing if disconnected from a mainframe. Then, in the late 1980s and the early 1990s, companies replaced their mainframes with networked PCs. The users of these new small computers gradually became familiar with the idea that their PCs, unlike the dumb terminals, could "think" for themselves. The power was on the desktop, not in the central mainframe computer.

. . . . .

With this new way of thinking about the individual PC came a new way of thinking about the network that linked PCs. Unlike the centralized network of the mainframe, this network connected all the other desktops in the office. People grew accustomed to sending messages and files to one another and to consulting databases elsewhere in the office. Gradually such local networks expanded: they began, for instance, to link PCs in the different branch offices of individual companies.

At much the same time, the computer became inexpensive enough to become an affordable household gadget. Indeed in one year—1994—sales of computers to homes in the United States even overtook sales of computers to businesses. Just as the PC spread from the office to the home, so now is the Internet spreading. The two trends are closely related.

Indeed, broadly speaking, countries with a large number of PCs per head tend to have a relatively high rate of Internet penetration and vice versa. In the United States, 40 percent of all homes have a PC, over half of them with a modem (which allows a PC to be connected to a telephone line); Europe's average is only half that. Not surprisingly, the United States has four times as many households on-line as does Europe. By contrast, in Asia, where sales of home PCs overtook sales in Europe in 1996, Internet use is rising faster than anywhere in the world.

An open standard
Launching a better telephone or a new television channel is easier than it used to be and more so in the United States than in any other part of the world. But change is still not a simple matter: both industries remain under the dominance of big companies that directly and substantially influence the pace and direction of innovation.

The Internet is different. Not only do companies use it to sell products from books to flowers, but it has become a huge market in its own right for hardware and software. An enormous number of companies, many of them small, most of them American, have sprung up to create networking software. All aspire, one day, to be another Netscape.

These "Netpreneurs" are often people who would have found it difficult, twenty years ago, to find an entrepreneurial outlet for their skills: the cream of young graduates, often with mathematics or science qual-

. . . . .

ifications. Not only is this new market changing the atmosphere in graduate schools—who wants to admit these days that they do not have a business plan on the go, as well as a thesis?—but it is also accelerating the development of new ways to use the Internet. The basic reason: the Net entrepreneurs do not have to apply for a license or approach big companies for permission. They just have to find the money and the market.

Outside the United States, such companies are dismally rare. An exception, of a sort, is Business Objects, a French software company famous mainly for being utterly un-French. Not only does it have an Anglo-Saxon name and Stanford-graduate as founder, but it initially achieved a thoroughly un-French rate of corporate growth, from a $1 million venture-capital start-up in 1990 to a value of almost $1 billion by the start of 1996.

The new breed of Net entrepreneurs, attracted by this open marketplace, have brought with them new business models. The classic example is Doom, a gory computer game launched in 1993 by a company called id Software. Made available free on the Internet, the first level of the game was downloaded by fifteen million people. Its reputation spread on the Internet, where players, hooked on its thrills, willingly paid for further levels of complexity. The following year, id Software released Doom II through ordinary stores. Its founders now drive Ferraris.[14]

The freebie model was repeated by Netscape. It grabbed the early lead in the browser market using two ploys. First, Netscape's software was distributed on-line, rather than through retail stores, thus saving the costs of packaging and distribution. Second, by giving its basic product away free, the company kick-started the market. In six months, six million copies of the Netscape Navigator were downloaded, producing enough publicity for Netscape to start charging—and to distribute fifty million browsers in a mere two years. That boosted the market for Netscape's expensive "server" software, which companies use on their host computers to communicate with browsers.

The give-away model makes eminent sense when it costs nothing extra to make more copies of a product. Computer games and browser software cost money to develop but not to manufacture. Distribute them on-line, and the publicity also comes free. And, of course, publicity for the product also attracts more users to the Internet.

Table 4-1   Growth of On-line Service Users

On-line Service	As of Dec. 31, 1995	As of Dec. 31, 1996
America Online	4,500,000	7,700,000
CompuServe	4,000,000	5,300,000
Microsoft Network	600,000	1,800,000
Prodigy	1,600,000	900,000

Source: *Seidman's Online Insider*

On-line services

Having bought a home PC, many people then wondered what on earth
to do with it, apart from playing games. One answer emerged as
modems, which allow computers to connect with telephone lines,
became more powerful and were integrated into PCs. Companies such
as CompuServe set up services that allowed individuals to send messages to one another from their home PCs, just as they could in the
office, and to tap into electronic discussion groups and libraries of
information. In the 1980s and early 1990s, such on-line networks were
closed: users could communicate only with other users, and the service
providers were able to control what was available.[15]

With the growth of the World Wide Web, these two worlds began to
merge. Proprietary on-line services such as America Online, CompuServe,
Prodigy, and Microsoft Network (MSN, launched in 1995) had given many
people their first taste of the potential of networked computers (see Table
4-1). But once the World Wide Web made the Internet more approachable,
the advantages of on-line services became less obvious. Most of the good-
ies they could once offer exclusively to their subscribers—magazines and
news services, for instance—are easily available on the Internet.

By 1996, the on-line service companies offered low-price access to
the Internet as well as their other services. That, in turn, encouraged
even more people to experiment with the Internet. In the fall of 1996,
more than half of all home users of the Internet reached it via an on-
line service provider. But these companies continued to lose ground—
in relative terms—to the more basic (and less expensive) services of
ISPs, which mainly offer a connection in exchange for a subscription:
no fancy billboards or chat rooms or library services.

To survive, America Online offered its "all-you-can-eat" package. But
a safer route to survival will be to become something different: perhaps

. . . . .

"branded communities," to use Bill Gates's phrase, that offer subscribers a guiding hand through the maze of the Internet. To lure such
communities, MSN offers six "channels" aimed at six different audiences: women, children, teenagers, and so on, each with targeted
shows that run for a few weeks at a time. MSN is in effect using the
television entertainment model and applying it to Internet use.

What the Internet Does Best
. . . . .

The Internet has flourished because it offers people services different
from and better than anything available by other means. It combines
the computer's extraordinary power to process and store lots of information at low cost with the telephone network's ability to reach millions of people around the globe. Its two main uses have been electronic
mail and finding information. But now other uses are growing fast.
Indeed, the business model that will ultimately work best for the Internet may not be the current confusing mass of innumerable Web sites
and the search tools to navigate among them but, rather, variants on
two entirely familiar themes: the telephone and cable television.

Electronic mail

E-mail has been easily the most useful and popular Internet service. It
is almost as old as the Internet itself: the first electronic mail was sent
between two machines in 1972 by a BBN engineer called Ray Tomlinson. Hunting for a way to separate the name of the user from the
machine the user was on, Tomlinson hit on the @ sign—thus, in the
words of Katie Hafner and Matthew Lyon, the historians of the Internet, "creating an icon for the wired world."

Electronic mail has remained the Internet's most widely used feature. Perhaps 200 million e-mail messages now fly through cyberspace
every day.[16] By one estimate, two-thirds of those who use the Internet
for e-mail at work do not use any other feature.[17] Between early 1997
and 1999, the number of electronic mail boxes is forecast to double,
from 75 to 150 million.[18]

Coupling the ability to send a message with the ease of perfect digital reproduction, electronic mail makes it virtually costless to send many copies to many people. It is thus ideal for distributing what would otherwise be heavy lumps of paper, such as office telephone directories or standard business forms. It is also, sadly, ideal for the campaigner, the junk-mailer, and the crackpot. Increasingly, people will probably use two electronic mailboxes—one public and one private—to keep such undesirables at bay. Because it is effectively free, it is depressingly prone to overuse. One chief financial officer of a large Silicon Valley firm recalls returning after a week's absence to find two thousand e-mails waiting. In despair, he deleted the lot, unread.[19]

Even so, many companies make increasing use of electronic mail. Instead of mailing out reports, companies will put them on the Web and send relevant staff or customers an e-mail with a link to the site. A manager who wants to send a detailed drawing to a colleague on the other side of the world will simply attach it to an e-mail. The result will be less expensive and of better quality than a fax.

In addition, electronic mail will serve numerous common uses:

- Sending out forms. Loan applicants and tax payers, will download the form, fill it in on-screen, click on hypertext links for additional information and answers to questions, and then click to submit the completed form.

- Paying electronic bills. Bills will arrive electronically, and the customer will type in a credit card number and click to send the payment off.

- Keeping families in touch. Instead of sending out pictures of a new baby, proud parents will attach the snapshots of and details about the youngster to an e-mail sent to a distribution list of relatives.

Finding information

One of the biggest revolutions made possible by the Internet has been access to information. By slashing the costs of this, the Internet brings together previously uncollated material and makes affordable material previously too expensive to obtain. It also, of course, allows publication of screens full of worthless dross. But overwhelmingly, the Internet allows users to become their own librarians, able to research, study, and investigate anything with nothing more than a mouse and a keyboard.

Most of the available information is free. Some has been put up by fame seekers, some by altruists, and some by advertisers. The Gutenberg project, for example, consists of a group of volunteer academics who put on the Internet out-of-copyright books, such as the works of Shakespeare. Some authors and many public bodies also voluntarily put information on the Internet.

Much of this information has always been in the public domain, but it was previously inaccessible to most people—because it was held in some special place or released only to certain groups of specialists. At one time, a city official might have made a policy decision known by sending out a press release to a few journalists and posting it on a bulletin board at the town hall; a few local people might read about it several days later in the local newspaper or happen upon it when visiting city offices for other purposes. Now that same decision can be published instantly and read by everybody with Internet access.

But there is a catch: "The Net for the first time is causing information overload," as Marc Andreessen of Netscape puts it.[20] Too much information may be less of a problem than too little, but it is still a problem. Many new Internet businesses aim at helping companies shout above the hubbub—or at helping people filter out unwanted noise.

Faced with a tidal wave of information, Internet users risk drowning. Learning search skills will become essential. Computers do not (yet) employ the sort of fuzzy logic that characterizes that most marvelous of all search tools, the human memory. Internet users today must, therefore, find the right word or phrase to reach the information for which they are searching. Bad spelling, acronyms, synonyms, and words that mean several different things can easily throw search systems into confusion. Authors with unusual names have people ringing them up to say their books are not mentioned on the Internet—only to discover that they were spelling the authors' names wrong.

Several companies have created guides to help users navigate the Internet, even if they do not know quite where they want to go.[21] Such guides make money by selling advertising, displayed on-screen along with the guide's index. As the popular guides are consulted up to fourteen million times per day, their computerized billboards are seen by lots of viewers.

The best-known index, Yahoo!, sorts sites into categories, just as a libraries sort books. It suffers from problems familiar to librarians: is this book, for example, about business or about communications? Search engines, such as Lycos, InfoSeek, and AltaVista, work on a different principle: they send software "spiders" out into the Web to follow every link on every page. When search engine users type in a few key words, the engines return every Web page that includes those words. Unless those key words are chosen with skill and an understanding of how the search engine works, however, the user will often be deluged with information.

Search engines are not yet good at distinguishing between, say, a press release and a detailed news report. They find it difficult to know which page will most interest a particular user. New kinds of software are being developed to try to solve this problem. By learning user's interests and tastes, they will be better able to target the most appropriate sources of information. These new programs, too, will know more about English, so that users can put their questions in a more natural way.

Telephony

Carried mainly over networks leased from telephone companies, using the Internet costs a minute fraction of the price of an international telephone call. This price gap is an invitation to arbitrage.

The market for voice telephone services is worth some $600 billion worldwide. If the Internet grabbed only a small part of that immense business, its value would be transformed. For a long time, the sound quality was pre-Marconi, and users needed to be waiting at each end with a computer plugged in and a sound card in place. But in 1996, that began to change, when IDT, a New Jersey call-back company, launched the first commercial Internet service to ring another person's telephone. The sound was still dreadful, making a callers sound as though they had a cold and a bad stutter, but that may be a small sacrifice in exchange for calling the United States from abroad for a mere ten cents per minute. More services, with better sound, are appearing.

. . . . .

Radio

As sound and video transmission improve, the Internet becomes an extraordinarily inexpensive way to distribute radio programs. For example, Progressive Networks, a Seattle company, has written software called "RealAudio," which allows users to listen to sound as it is transmitted, rather than waiting to download it first. As a result, a user can run a radio station on the World Wide Web that works just like a broadcast station, but at much lower cost. Radio stations emanating from one point on the globe can thus be available to listeners anywhere else on the planet. A transmission can be stored to be listened to later—creating radio-on-demand. The snag: because of radio's demand for capacity, a couple of hundred listeners can easily jam access to an on-line "station."

Personalized broadcasting

Could the Internet become a low-cost way to deliver a new sort of cable television? "Webcasting"—putting live video pictures on the World Wide Web—is, in Internet years, old stuff: a Rolling Stones concert was transmitted over the Internet in 1994. But it has moved on. In 1996, many companies began experimenting with ways to "broadcast" content continuously to users' computers.

Webcasting works much like electronic mail. Instead of having to search for information—to "visit" a site and "pull" the information out—users sit back and relax while special services "push" information or entertainment onto their computers. PointCast, one of the first examples, regularly pumps news headlines and other information onto a screen-saver on users' PCs, much like a ticker tape on a cable-television screen. Users can follow particular subjects they are interested in, with news flashes that trail across the screen.

The touted advantage of such services is that they obviate the need to wrestle with search machines. "Manually searching the Web is not a sustainable model, long term," announced Eric Schmidt, former chief technologist at Sun Microsystems and now head of Novell. The most widely used sites on the Internet are those that help people to find their way around cyberspace—a clear sign that people are lost. Radio and television, so goes the theory behind PointCast and similar services, are simpler, more familiar mediums than the tangle that the World Wide Web has become: they need less technical expertise to use and, sponsored by advertisers, offer plenty of free content.

But this model of the Internet has three catches, compared with the old one. A single broadcast channel, scrolling across the screen, may be tailored to the user's interests, but it gives the broadcaster powerful leverage. After all, while the "pull" model allows the user to visit any one of thousands of sites, the "push" model gives one company control of the path to your desktop. Broadcast also works best with a continuous Internet connection, such as most people have at work—but few people have at home. The technology thus widens the existing gap between office and home users. Finally, companies may not be happy about yet another distraction vying for their employees' attention. After all, companies don't pay their staff to watch TV; why should they pay them to watch the Internet equivalent? Some companies, such as Hewlett-Packard, have as a consquence tried to discourage the use of networks such as PointCast.

Like the Web, the main use of this sort of Webcasting will probably turn out to be corporate. Some companies already use in-house broadcasts to distribute company information—or simply to feed software programs into employees' computers, ready for use at another time. The broadcasts often run over corporate intranets: for example, Ben & Jerry's, a fashionable American ice-cream marker, has devised a Webcast system to inform managers automatically when the company's reserves of ice cream fall below designated levels.[22]

Obstacles to Growth
· · · · ·

Before the Internet can become a true mass medium, rivaling the telephone and television, two problems must be resolved. First, users need access: individuals must be able to use Internet services from home. They will need a PC or something similar and an adequate home connection. Second, the supply of quality service must meet the demand of those companies that want to use the Internet to sell their wares. The Internet needs to be more secure and less congested.

. . . . .

Access: Connecting the home

Many people who use the Internet at work find the low quality of the connections available from home enormously frustrating. Dialing up a local ISP not only means a much slower link, so that information can take forever to download; it also means that, in order to taste fully the wonders of Webcasting, the line must be left permanently open.

Upgrading the connection with fiber-optic cable to the curb and coaxial cable to the home, coupled with digital switching and transmission, may eventually allow home Internet access as good as most people enjoy at the office. In 1996, cable-television companies in several countries launched high-speed access to the Internet on their networks. But there are problems. In most countries, far fewer people have cable television than a telephone. In addition, the arithmetic is off-putting. A typical estimate of the cost of upgrading existing telephone service to receive video is $1,000 per home. Companies are making those investments— but only where they can see an average return per home passed of, say, at least $200 per year. Some households, of course, would not make use of any of the new services that such a network offered, so the cost of running a connection past their front doors would have to be recouped through higher spending by their neighbors. If only half the households passed took the service, then its builders would need to be confident that they could make $400 per year from each subscriber.

Such figures are not unimaginable; after all, the average American home spends roughly that on a cable-television subscription and more than twice as much on the telephone each year. But both of these are well-established services. The Internet might require many years for revenues to build up to such levels. And for the infrastructure investment to pay off, investors must be assured that people (or advertisers) will pay such sums over and above what they already pay for television or telephone calls. As a result, upgrading will happen in some areas (such as prosperous university towns) much faster than in others (such as crumbling downtowns). The wealthy will be on high-quality Internet connections from home long before anybody else.

Companies will develop halfway measures that are less expensive and risky, sometimes combining two different delivery channels. DirecPC, a subsidiary of Hughes, offers Internet material one-way over a satellite link (people use a telephone line as their return path); and telephone companies all over the world are working on digital compression, to squeeze more through their narrow copper wires. But the bottom line remains: until the early years of the next century, and in many places for even longer, Internet access will not be as good at home as it is at work.

Access: The PC

In the United States, one home in five now has a PC with a modem and so can at least have dial-up access to the Internet. Two homes in five have PCs, and it is a fair bet that many of those will upgrade over the next two or three years, buying a model with a built-in modem. But what about the remaining three out of five?

Some of them, no doubt, will acquire PCs—especially if they have children: homes with young families are much more likely to acquire a PC than those without. But the PC is still expensive compared with, say, a television set or a videocassette recorder, and the price, at least in the United States, has shown no sign of falling. (In Japan, where fierce competition caused PC prices to halve between 1995 and 1996, the result was a 70 percent rise in PC penetration, followed by a sharp rise in Internet connections.) But generally speaking, if the gateway to the World Wide Web remains acquisition of a PC at a minimum cost of around $1,500, its services will never be as widely used as those that can be delivered via the television, the radio, or the telephone. If the Internet is to penetrate more than half the population in the rich countries, or to reach the levels in developing countries that it has in the United States, new access technologies will have to be devised.

Might the gateway to the Internet turn out to be some other less expensive machine, such as a digital set-top box, WebTV, "network computer," or something similar? Until it is clear what services will be required, it will not be obvious what sort of machine would be most appropriate as a receiving device. After all, the two existing household screens—the television and the PC—are used quite differently and in different parts of the home. The television still sits, in many homes, in a communal room; the PC is more often in a private one. People use their television sets differently from the way they use their PCs: they sit

· · · · ·

(or slouch) several feet away, using a remote control rather than a complex keyboard or even a mouse.

The driving force here may turn out to be digital television. After all, if people are acquiring digital televisions, in effect a home computer, often bought at a price subsidized by a satellite-broadcasting company or hired from a cable company, these may as well incorporate Internet access. The convergence of the Web with broadcasting may also make that more likely.

But if the PC remains the principal gateway to the Internet, then another natural limit to Internet growth imposes itself. At present, there are two hundred million PCs, at most, worldwide (probably many fewer). This number grows by about 15 percent per year, a far slower rate than the Internet's growth of more than 100 percent per year. Without the development of new items to connect, such as television sets or network computers, growth of the Internet will inevitably slow.

Quality of service

For companies hoping to sell services over the Internet, its two biggest limitations are security and congestion. Valuable commodities that can be digitally transmitted, whether money or sensitive information, will be entrusted to the Internet only if they can be safely encrypted. And the Internet will not be useful to companies if it takes too long for information to appear on even the fanciest screen with the largest connection.

Security

The main reason for the Internet's success, charm, and enormous attraction for innovators is that it is an open network across which anybody can send any information. But this openness also creates headaches for those who want to use the Internet for valuable transactions.

Broadly speaking, the system is vulnerable to damage from two sources: hackers and viruses. Both use the Internet as an entry point into a company's computer network, from which vantage they can intercept in transit material such as credit-card numbers or private information.

The threat from hackers and the costs of keeping them at bay are both growing rapidly. Several factors increase the security risk they present.

First, thousands of proprietary systems, which would once have been understood only by a handful of experts within a company, are being replaced by the universal currency of the Internet protocol. Second, individual employees are acquiring more power to sabotage: sensitive information about customers can be copied without trace and electronically distributed. Third, the sheer amount of damage a hacker can do has risen as industries have become more and more dependent on their computer systems. With larger systems comes ever greater potential for mutilating taxation and social-security records, disrupting air-traffic control, interrupting electricity supplies, or stealing large sums from banks.

Security has not kept pace with the growth of the Internet. A 1996 survey of Internet sites found that more than 30 percent had no "firewall," a program to prevent intrusion; the same survey revealed that more than half of the U.S. information security managers consulted had experienced Internet-related security breaches in the previous year.[23] Some estimates suggest that every twenty seconds a computer on the Internet is attacked. In the most dramatic larceny case, a group of Russian hackers transferred $10 million out of Citibank's cash-management network in 1994. Working with a private security firm, the bank eventually recovered all but $400,000.

Nobody knows how much hacking takes place. (One problem with estimating the extent of both hacking and viruses is that statistics on their occurrence of these problems comes mainly from companies trying to sell protection equipment; depending on their numbers is rather like depending on a locksmith for data on area robberies.) Bruce Scheier, author of a book called *E-mail Security*, maintains, with a certain hyperbole, that "The only secure computer is one that's turned off, locked in a safe, and buried 20 feet down in a secret location—and I'm not completely confident of that one either."[24] But most attacks—95 percent, according to America's Federal Bureau of Investigation—go undetected, and of those that are detected, only about 15 percent are reported.

Computer viruses—rogue programs that can disable a computer or destroy data—also seem to be a growing problem. Some viruses are planted deliberately and maliciously by hackers; more often, they are introduced unwittingly by a user. Virus infections grew almost tenfold in American companies in 1995.[25] Some people worry that the rise of Webcasting may spread viruses even faster, because many of these programs automatically download software into users' PCs so that it can sit there, ready for use. "If it's buggy, you've just hosed down a million people," says Novell's Eric Schmidt.[26]

Interception
The Internet has been an open, public network for most of its life. But commercial use requires privacy and security, two qualities that are difficult to incorporate into an open network without limiting its growth. The necessary trade-off may ultimately be made through encryption.

Encryption can protect material traveling across the Internet, whether a private message, a purchase order, or a credit-card number. Encrypted information—information that is scrambled, so that only somebody with a secret key to the code can read it—can travel across an open channel or be stored on a hard disk, concealed from those without the necessary key. Encryption thus gives protection both in transit and in storage.

The development of robust encryption has caused the American government years of soul-searching: after all, if encryption could be used to prevent hackers from stealing a credit-card number or a software program, it could also be used by criminals to send plans and money around the world. Concerned that encryption might be misused, the U.S. government has fought a long battle to prevent the American public—or, worse, foreign citizens—from being able to use encryption technologies that U.S. government agencies could not easily decode.

Encryption, together with authentication systems to ensure that documents sent over a computer network have indeed been sent by the person purportedly sending them, will need to become standard features of all electronic business transactions. If governments do not allow that to happen, the growth of electronic trade will be stunted. But even if they do, companies will spend huge amounts of money and management time fighting to protect their security and that of their customers.

Congestion
In its early days the Internet carried data mainly in the form of text; now, increasingly, it carries services that need far more capacity. This new demand has led to congestion and crashes. Expanding capacity is one answer to the problem. But in addition, the Internet needs to

develop a more refined pricing structure. Without one, it will not be able to offer the quality of service that commercial users require.

The delays on the Internet have several causes. Some of the most popular sites simply do not have large enough connections to handle all the demand from users: perhaps 90 percent of traffic on the Web's busiest links is destined for forty or fifty popular destinations.[27] The jam that occurred at America Online in early 1997 resulted from switching all subscribers to flat-rate pricing, creating an invitation to hog the line without, at the same time, putting in the equipment necessary to handle the increased number and length of calls.

Some local telephone companies find Internet users clog their lines. A system of flat-rate pricing for local calls works with the telephone, where time imposes a limit on how long people want to talk, but not with an Internet connection, which can be left switched on all day. In the United States, where the average length of telephone use is almost half an hour per day (roughly three times longer than in Europe), heavy Internet users may remain on the line for an hour and a half per day.[28]

The sheer volume of Internet use has risen faster than many parts of the network can cope with. Traffic through one of the main connection points doubles every six months, while the number of Internet subscribers doubles only every year.[29]

The jams may well get worse as new bandwidth-gobbling applications expand. In mid-1995, the World Wide Web already accounted for more than one-third of all the traffic on the Internet; by early 1997, it was closer to two-thirds.[30] Internet telephony and radio exacerbate the problem, but the true culprit will be video, the ultimate congestant: one million bytes of capacity can hold the text of a 700-page book but only fifty spoken words, five medium-sized pictures, or a mere three seconds of high-quality video.[31]

With improvement in the capacity and connection quality to homes, the problem of congestion may worsen. A charge for quality service will be the likely result. Such charges would have two effects: first, they would create an incentive to improve the infrastructure and thus to increase capacity; and second, they would persuade users to be more frugal.

A tier of prices might then emerge, rather like the one suggested above for telecommunications (see Chapter 2), with higher charges for users wanting guaranteed services and lower charges for everybody else.

It will simply not be possible to guarantee minimum levels of service without charging a fee. An effectively free Internet will be rationed by congestion. Users expect an Internet connection to work as quickly and effortlessly as changing the channels on their television set or calling a service on their telephone. If congestion becomes severe and this standard is not met, those who need the Internet for work will stop using it.

To provide guaranteed service, companies—probably the telephone companies—will build private networks. These will carry the more valuable traffic, guaranteeing priority, service, and security—in return for a fee. These alternative networks are already starting to emerge. In fall 1996, a group of American research universities announced a plan to build Internet II, dedicated to academic traffic and free of commercial users, just like the early Internet.

The development of an effective way to price the Internet will be crucial to its future. If that can be done, the important question will be whether its growth slows. How far, in other words, has the Internet's phenomenal expansion been due to its users' perception that it is virtually free?

Why the Internet Matters

. . . . .

Nobody can tell what the Internet will become—it is simply changing too fast. Internet broadcasting, for example, went from a curiosity to the cover of *Business Week* in a few short months. Clearly, the Internet matters. The three main reasons are its global span; its ability to merge the capabilities of television and the telephone and to move beyond them; and its stimulus to innovation.

Global span

Because more than half the users of the Internet are in one country, its role as a global medium is barely apparent. The telephone still connects continents much more effectively than does the Internet. But that is changing at tremendous speed. The fastest growth rates for Internet use are outside the United States. In 1996, the number of Internet hosts in Japan grew by 173 percent and in Hong Kong by 178 percent. The newly industrialized countries of Asia, with their passion for gadgets, their advanced electronics industries, and their young, well-educated populations, will be the perfect Internet market.

Poorer developing countries will be further behind, simply because of the expense of high-capacity networks. But even for their citizens, the Internet offers freedoms: a way around overpriced international telephone and postal services, for instance, and a short-cut to information that may be unavailable locally, such as scientific articles and unbiased news.

Once it has true global reach, the Internet may well become the main platform for international contact. It will provide, for instance, a shop window in which a company in one country can display its wares to a world market. It will offer a chance for people in different countries to swap information and ideas. It will provide the means for people in countries in which the main news media are censored to tell their stories to the outside world and to contact sympathizers. No other innovation has ever had such earth-shrinking potential.

Convergence

The Internet is neither telephone nor television, but it shares some of the characteristics of each. More important, it is changing both and will in the future partly merge into them.

In the case of the telephone, the Internet's most dramatic impact will be on rates. As Internet telephony develops, so it will change the basis of call pricing, replacing tariffs for distance and duration with fees based on quality of service. In addition, the Internet provides a way to try out new combinations of telephone service. (The videophone, which failed as a telephone product, is more successful on the Internet, for example.)

In the case of television, the Internet offers a bridge between the continuous flow of information (the television model) and the focused search for facts (the Web site model). The Internet will drive the uses of television in the work place: this venue has never much interested the big broadcasters, but it is the perfect outlet for Internet television.

Eventually, the Internet will be integrated into other products. It will be part of the telephone service, part of the way a television works, part of a games console. It will connect computers in an office and computers between offices. People will stop thinking of the Internet as a separate entity and will be aware only of the services it delivers rather than the technology itself.

Innovation

Most important of all, the Internet has become the most powerful driver for innovation that the world has ever seen. Because of its open, flexible protocol, thousands of small companies, founded by the best-educated group of entrepreneurs ever to blitz a business, are making (or, periodically, losing) huge sums of money developing new ways of using the Internet. The same phenomenon appears in other countries, paralleling their accelerated use of the Internet.[32]

The result has been partly to change the structure of the communications industry, shifting the focus of innovation away from the old giants towards these young hothouses, but even more importantly, it has been to drive forward communications technology at a formidable pace. Through the Internet, new products can be relatively inexpensively developed and launched; potential customers and investors can be targeted; and markets can be quickly identified and tested. No other industry has such instant links between inventors and their customers.

That, above all, is why the Internet matters: it makes the market system—since the collapse of communism the dominant method of allocating resources around the world—work better. Those parts of the world that embrace the Internet will find themselves better able to compete than those that lag behind.

5

Commerce and Companies

The changes sweeping through electronic communications will transform the world's economies, politics, and societies—but they will first transform companies. They will alter the ways companies reach their customers, affecting advertising, shopping, distribution, and so on; they will create new businesses; and they will change the way companies communicate with one another and with their staffs.

Companies will be affected first partly because they are best prepared in terms of equipment. Connections to businesses are generally of better quality than those to homes. Because businesses tend to be clustered in districts, their connections can be upgraded relatively inexpensively; and, because they make more use of communications, they generally value their connections more and so are willing to pay more for them. Not surprisingly, business districts have been the first places to get the high-quality fiber-optic networks that make possible fast and easy connections to other computers. Even in the United States, the quality of service available to domestic users of the Internet will not begin to match that of corporate ones until the turn of the century; comparable improvements will be later still in most other countries.

But in addition, the changes in communication chime with other trends in the corporate world and will reinforce them. One such factor is globalization, with world trade and investment accounting for a rising share of total economic activity. Another is the booming market for ideas. Increasingly, creativity and the ability to process information will be key competitive tools. In addition, businesses must respond to the incessant pressure to control costs. With the return to an era of low inflation, reducing costs has become a survival tactic essential for many companies. Related to this is the impetus at many companies to cut layers. Companies increasingly need to remove internal layers (of middle management, for instance) and external layers (intermediaries). Flexible employment patterns, designed to acquire the maximum expertise at minimum cost, create different relations between companies and their workers and require flexible communications.

Adept use of communications will clearly become an important—perhaps the most important—competitive advantage for businesses. It offers a chance to create new business models and to find entirely new ways of doing things, and it provides opportunities for building new relationships with existing customers and for tapping markets that were previously hard to reach.

In this new game, American companies have already acquired a lead that should dismay their competitors elsewhere. Year after year, the United States has been spending a larger share of its GDP on computer hardware and telecommunications equipment than has any other country. U.S. investment in communications, as a share of national output, is roughly double that in Japan. (See Figure 5-1.) The United States still overwhelmingly dominates use of the Internet. Venture capital investment, the primary source of finance for the many tiny companies that have sprung up to exploit innovations in the communications and computer industries, is far more available in the United States than in continental Europe, let alone Japan.[1]

Not surprisingly, therefore, almost all the innovation in corporate uses of new communications comes from the United States. These innovations are transforming advertising, commerce, corporate structures, links with employees, and corporate size. For companies everywhere, the best guide to the future will be watching the way these changes sweep through the U.S. corporate world.

Advertising

· · · · ·

Advertising attracts consumer attention, and communication has always been at the center of the industry. The introduction of television advertising fifty years ago brought about a revolution in advertising. Suddenly, companies could reach customers in their homes, simultaneously and nationally, and more vividly than had been possible with the

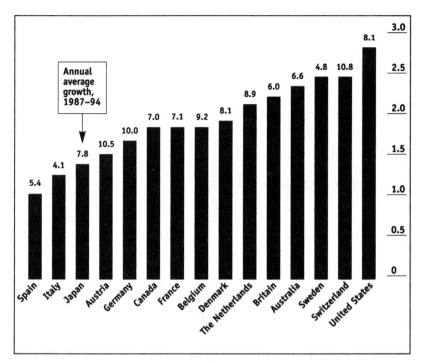

Figure 5-1   The Growth of Information Technology
Computers and telecommunications spending as a percentage of GDP*, 1994.
*Excludes software and services.
Source: ©*The Economist*, London (November 23, 1996).

printed page or the radio. The effect was to transform retailing and to boost the power of brands.

Brands, the stock-in-trade of advertising, are the creations of national communications networks: Ivory Soap was first floated in 1879, American Tobacco was founded in 1881, Coca-Cola and Johnson & Johnson were both founded in 1886—products, all of them, of the heyday of the telegraph and the railroad.[2] The rise of network television gave new strength to brands, helping to shift the balance of power from stores to manufacturers: it is difficult now to imagine Procter & Gamble, the global giant of the household-goods market, without television advertising.

But communications are changing again, in ways that will reshape advertising and hence advertisers. The multiplication of television channels fragments audiences, and the Internet offers new advertising models, although its audiences are still small. Overall, advertisers will have many more choices of outlets for advertising—but the audiences in each outlet will be more specialized. Advertisers will have more potential for global branding—but markets will remain culturally distinct. Changes in communications will increase what companies know about their customers, will fragment advertising markets, and will shift company relations with customers from passive to active.

Information about customers

In the electronic world, people leave a lot of tracks. Stores know what customers buy from their preferred-customer cards ("loyalty" cards, as they are called in Britain); from their credit-card transaction records, banks know where customers spend their money; mail-order firms know where their customers live; airlines practically know their passengers' deepest thoughts from the personal information they hand over in exchange for being allowed to join frequent-flyer programs.

The communications industry in particular knows more and more about its users: telephone companies know who telephones where, and, when people use pay-per-view, television companies know who watches and when. Companies with Internet sites know who has visited and what they wanted to look at, rather than just how many people stopped by. In the future, one of the main roles of communications companies will be managing the information they collect on the transactions and behavior patterns of their customers. Where communications companies bill customers, they have a particularly powerful tool for reaching and understanding their market.

All these electronic tracks delight advertisers, to whom they convey the means for analyzing customer data from every conceivable viewpoint. Once advertisers learn how to manage the deluge of data, they will find that targeting individual customers precisely will become—or should become—much easier. The result will be a growth of "below-the-line" advertising: direct mail and its electronic equivalents, telemarketing, and e-mail. The new "push" techniques of Internet broadcasting offer a perfect platform.

Indeed, the Internet allows companies to target individual customers more precisely than has been possible in the past. Customers who search for information on buying a car or a CD can be led seamlessly and with increasing precision to an advertiser's site or to a list of possible purchases. And companies will be able to measure, more accurately than ever before, how many people are sufficiently interested in their advertisement to follow it up. In 1996, Procter & Gamble refused to pay for advertisements that did not attract enough clickers.[3]

The ideal, however, may be better than the reality. The electronic tracks may lead, with annoying frequency, to a relatively small group of potential customers: the well-off and the young. People may increasingly resent being singled out as targets for advertising. If they do, they may find it hard to drive persistent companies away: it can take weeks or even months to remove one's name from the hit list of the telephone sales staff at a large American long-distance carrier such as AT&T, for example. Customers may start hunting for ways to seal up the electronic mailbox.

Fragmentation of advertising markets

As television channels proliferate and Web sites multiply, large audiences become harder to find. The declining share of mass-market television means declining opportunities for mass-market advertisers. This fragmentation has many effects, including increasing the importance of well-known brand names and boosting the demand for celebrities with a mass-market reputation to help to sell products.

On the Internet, large audiences are particularly difficult to reach. The World Wide Web seems at first glance an advertiser's dream: allowing access to a Web site costs much less than printing and distributing a brochure; updating or customizing is quick and simple; and as many as fifty-seven million well-heeled and literate shoppers might look at it. (See Figure 5-2.)

For the moment, these advantages remain largely theoretical. Well into the first decade of the next century, the Internet will have a diminutive slice of total advertising revenue. In 1996, global spending was around $260 million,[4] a flea-bite compared with the total of $173 billion spent on advertising in the United States in 1996. American cable television alone receives $6.6 billion per year in advertising, a figure reached only after a decade and a half. Moreover, more than half of Internet advertising is by American software and computer-service

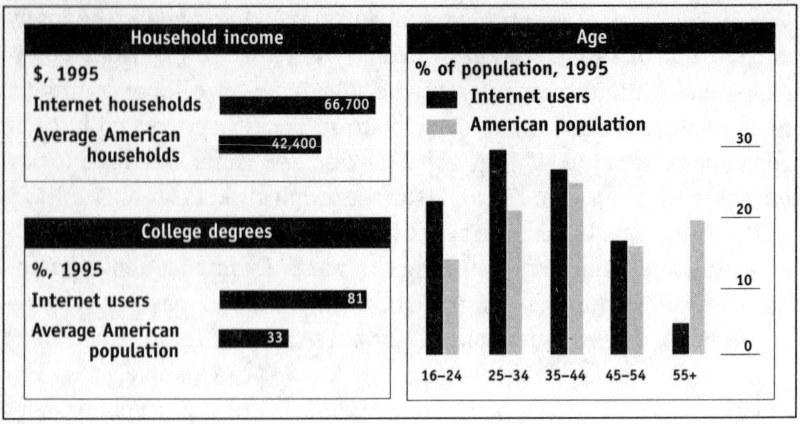

Figure 5-2 Advantages of the Internet
Source: ©*The Economist*, London (October 26, 1996).

companies (led, unsurprisingly, by Microsoft). As many of these also own the Internet's prime advertising sites, much advertising space is merely bartered. The Internet seems an eminently suitable advertising medium for automobile companies and financial services firms, but they account for just over 5 percent of cyberspending on advertising.

But that will change. Some advertisers already see the Internet as an inexpensive place to experiment. London International, a British company that makes condoms, is using its Web site to experiment with advertising pitches before transferring them to print and television, in the hope of avoiding expensive mistakes.

In the future, companies may well find that the task of reaching niche markets becomes relatively easier, compared with that of reaching mass audiences. If so, the result will be a growth in niche players and products. That will chime in with a trend in the market at large: the desire of consumers for more customized goods and services.

A shift from passive to active advertising

Advertisers will be affected by the fact that ways to keep unwanted material off the screen will grow more sophisticated: the technologies that prevent children from seeking pornographic Web sites or watching violent television shows will be adapted to keep out advertising, too, if that is what consumers want. So the trick will be to make customers want to watch a marketing pitch—because it is fun or informative or because they get something else in exchange. Advertisers will need to convince consumers, in Bill Gates's phrase, to "invite Web advertising into their lives." The result will be the growth of what Toby Braun, an American advertising executive, once called "bartered sponsorship": you watch this commercial, and we will offer you free e-mail or show you this movie for half the price you would otherwise pay.[5]

Internet users are constantly offered access to desirable information (such as a newspaper library or a sports service) in exchange for giving their names and addresses and information about their income and tastes. A more direct bribe, announced in January 1997 by the Swedish company Gratistelefon, offers free telephone calls but interrupts them periodically with ten-second advertisements. A one-month trial in two Swedish towns generated about 30,000 calls per day. The caller dials the company's toll-free number and then the number of the person they wish to call. A commercial plays while the caller waits to be connected, at the end of the first minute, and then again once every three minutes. The high proportion of young callers attracted advertisers for snacks, a radio station, and a cinema chain.[6]

Classified advertising is an electronic natural. Users will find it easier to hunt for a secondhand Ford or a new job or apartment by clicking a mouse than by turning the pages of a newspaper, and advertisers may be willing to pay extra to have their ads appear in both the newspaper and on-line. In 1996, for example, Rupert Murdoch's Australian publishing business launched a Web site carrying classified ads from his large stable of newspapers (he owns about 60 percent of the market). Customers can search ads from individual newspapers and from national databases created by aggregating certain categories of ads. The idea is that customers still buy the newspapers to read the news, but advertisers willingly pay more to have both the print and the on-line display.

Building branded communities

In the struggle to attract attention, advertisers will find it easier to target individuals but harder to find mass audiences. But one of the most impor-

. . . . .

tant innovations of the Internet may be to allow companies to extend brands by creating a sense of community among customers using them.

The basic idea is to build links with customers, using the Internet's networking power. So, for instance, anybody who owns a Saturn car can register as part of an "extended family" of Saturn owners and discover the names of other Saturn owners living nearby. Other companies have created bulletin boards on which customers can tell each other what they love or hate about a particular product or service. This is a key concept behind Motley Fool, a personal-finance site that allows investors to swap tips. Another site, Firefly, specializes in trying to cater to individual tastes; its BigNote is an on-line music store that asks customers to rate a range of artists and albums. That may introduce them to new albums that they like.[7] Another example is Cobra Golf, an American golf-club manufacturer that has created an unedited bulletin board on its Web site for customers to exchange comments with one another about the company's products.[8]

This marketing tactic has all sorts of advantages. Turning one's customers into a virtual community offers a way to use the powerful economics of a network to reach like-minded customers—to build a brand, get feedback on products from users, and extend the corporate culture to include the customer. But it could also turn out to have disadvantages, if it increases the exposure of companies to consumer pressure groups, or if the chat is hijacked by lobbyists or by a few disgruntled customers. The shortcomings of a product may be magnified out of all proportion. From a company's point of view, turning customers into a community will be a mixed blessing.

Electronic Commerce

. . . . .

Improved communications will transform many aspects of commercial activity. Among its many consequences will be lower business entry costs; better customer information; easy price comparison; increased pressure on agents and intermediaries; lower distribution costs; global reach; and new payments techniques. Underlying all these changes are new ways to reach customers through the telephone and the Internet. Although the Internet will be the more novel and will stimulate most innovation, the telephone will remain the most important communications tool of commerce. Retailing and distance shopping will attract the most attention, but the most significant impact will be on business-to-business commerce. A single large company, GE, now does more electronic trade with its suppliers than all the retail commerce on the Internet combined.

The tools of the trade

The three different telecommunications technologies have different strengths and weaknesses as commercial tools. Television shopping, for example, tends to be best for moving large quantities of a few items (such as jewelry), while the typical Internet retailer sells small quantities of many items (such as books).

Telemarketing accounts for several hundred billion dollars per year of commercial transactions. The biggest spur to telephone commerce has been the toll-free number, which not only facilitates national advertising campaigns ("Dial 1-800-FLOWERS") but makes call-handling efficient. Incoming calls can be routed through a central point and then out to the geographically closest or the least busy branch for dispatch.

Toll-free calling is old news in the United States, where it was introduced by AT&T in 1967; in 1996 it accounted for $157.4 billion of commerce.[9] It is one of the many areas where the United States (followed fairly closely by Britain and Canada) has raced ahead of most other countries: American telephone companies carry around one hundred million toll-free calls per day to more than eight million 800 numbers. (See Figure 5-3.) In contrast, in 1995, Japan had a paltry 300,000 toll-free numbers, and Germany had half that number. The idea that a company should pay for its customers' calls still seems odd to many companies outside the United States.

In 1997, telemarketing took a great leap forward, thanks to the launch in the spring of global toll-free numbers. These enable companies to have a single toll-free number that can be used all over the world. No longer will mail-order companies, broadcasting advertisements by satellite television to hotels around Asia or Europe, have to end their ads with an interminable string of different numbers for different countries.

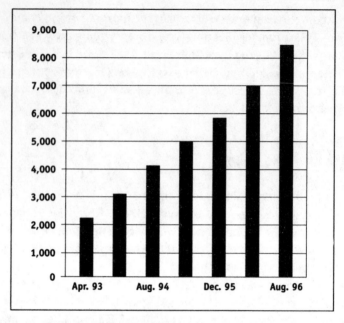

Figure 5-3   Growth of the Number of Toll-Free Telephone Numbers in the United States, 1993–1996 (in thousands)
Source: International Telecommunication Union "World Telecommunication Development Report, 1996/97."

The result will be a boon to all sorts of industries: hotels, airlines, credit-card companies, mail-order firms, vehicle-recovery services, and call-back companies. Companies that understand how to use toll-free numbers will increase their global reach. Long before the Internet becomes a widely used global marketplace, the telephone will be there. By binding global toll-free numbers to the Internet, customers looking at an on-screen advertisement will one day be able to click on a window for voice contact with an operator. Two tools of global commerce will become one.

Television marketing has traditionally catered to a narrow market, consisting almost entirely of middle-aged American housewives lured mainly by jewelry and women's clothing. Those products together account for nearly 60 percent of the revenues of QVC, one of America's two big shopping channels.[10] In 1996, a subsidiary of Tele-Communications Inc., a big American cable company, had a successful launch of a television shopping network in Japan. In many parts of the world, though, television shopping is unknown.

Television may be a great way to advertise, but it is a poor way to sell: people have to be willing to spend time watching and must buy on impulse as the product flashes across the screen. In addition, television shoppers almost always have to dial a number shown on screen in order to buy. With digital television, that may change, and people may buy more if all they need to do is press a button on the remote control. With channels targeted at motorists, gardeners, and sports fans, television shopping may evolve to a degree as a sort of screen-based version of the studio store.

In theory, the Internet ought to be a much more effective selling medium than television. It reaches a group of customers that have often proved elusive: wealthy and intelligent individuals, especially youngish men. It also offers a completely new way to approach catalogue shopping, dispensing with one of the main costs for companies selling by catalogue, paper, and postage.

Thousands of companies have therefore set up shop on the Internet. They range in size from Internet-only small fry to Wal-Mart, the world's largest retailer, which started offering products on the Internet in 1996. They include virtual supermarkets such as Peapod, which sells groceries in several American cities, and real ones, such as Tesco, which has an Internet ordering service for customers in parts of London. Computer software and hardware were the first goods widely available on the Internet, and are still the most commonly purchased. Next come travel services, such as airline tickets, and leisure goods, such as CDs and books. Pornographic material, harder to measure, is also a best-seller.

For the moment, on-line commerce is tiny. Many companies say that thousands of people look at their wares every day but few buy. Even companies that have succeeded with on-line retail marketing are still mainly minnows. Virtual Vineyards, which sells only on-line, was hailed as a success when it took in $225,000 of electronic sales in the first quarter of 1996.[11] But any successful high-street wine shop would expect to do better than that. Forrester, a consultancy in Cambridge, Massachusetts, estimates the value of sales of goods and services on-line at $518 million in 1996, roughly one-tenth of the revenue from television shopping, which in turn is roughly one-tenth of the $43 billion that individuals spent on catalogue shopping in the United States in 1995, which pales beside total American retailing revenues of around

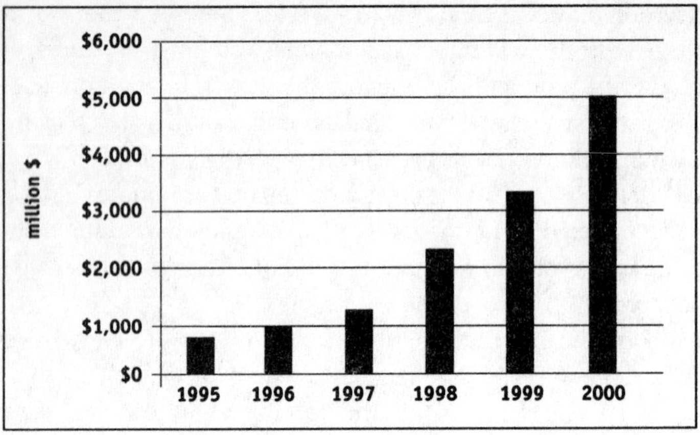

Figure 5-4   Electronic Marketplace Sales Growth, 1995–2000
Source: Cowles/Simba Information; (800) 307-2529.

$2,400 billion.[12] Few on-line retailers are known to be making money, and many are thought to be losing heavily. (See Figures 5-4 and 5-5.)

Distance shopping

How fast will distance shopping grow? Its prospects are certainly better in the United States than in any other country. But even in the Unted States, distance shopping has never accounted for more than a fraction of retailing. In 1980, nonstore retailing accounted for 2.4 percent of total retail trade. By 1994, that proportion had edged up to 2.9 percent, which represented a lot of extra spending. But the overall low percentage suggests that the vast majority of customers are perfectly happy to buy at real-world stores.[13]

The propensity to shop from a distance is a matter not only of cost and convenience but also of culture and the structure of taxation. In the United States, mail-order has been familiar since the nineteenth century, thanks to the Sears, Roebuck and Montgomery Ward catalogues. Americans are more likely to have credit cards, one essential tool for distance shopping, than people elsewhere. Express parcel-delivery services, the other essential, cost much less in America than over comparable distances in Europe. And America's tax system gives encouragement to dis-

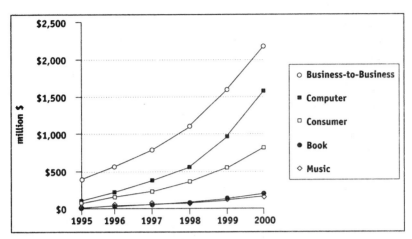

Figure 5-5   Electronic Marketplace Sales Growth by Category, 1995–2000
Source: Cowles/Simba Information; (800) 307-2529.

tance shopping: goods bought in one state but delivered in another escape sales tax.

No such concession applies to value-added tax, when goods bought in one European country are delivered in another. Instead, cross-border shopping in Europe runs into obstacles of language, standards, and regulations. Not surprisingly, distance shopping has flourished in America more than elsewhere. For instance, 39 percent of Americans buy books, CDs, or tapes from catalogues, but the comparable figure in France is a mere 20 percent, and in every other large European country it is lower still.[14] Television shopping barely exists in Europe, and telephone commerce, whether ordering a pizza, checking a bank balance, buying tickets, or hunting for information, is far more common in the United States than anywhere else. Where people are reluctant to use a telephone for a transaction, they are surely no more likely to use a keyboard with a screen.

While much discussion of on-line commerce has focused on the individual customer, the big markets, until well into the first decade of the next century, will be corporate. Most business transactions are already done at a distance—by telephone, fax, or mail. Companies generally have better communications facilities than do homes. And no corporate purchasing department regards an afternoon's shopping as a recreation.

Indeed, the biggest sales on the Internet are not earned by florists, wine merchants, book stores, or any of the other much-hyped on-line retailers. They are, rather, earned by companies selling mainly to other companies. A prime example, discussed later in this chapter, is GE; another is Cisco, the world's largest supplier of computer networking equipment, which probably accounted for 10 percent of all the business done on the Internet in 1996.[15]

It seems for the time being that computer shopping is a great idea—for people who want to buy products associated with computers. But that will gradually change.

Lower entry barriers

One of the Internet's most astonishing effects on commerce has been to reduce entry barriers. The Internet cuts costs in two main ways. First, it reduces the cost of marketing and of customer information. Anybody with the basic skills (or the ability to pay for them) can put up a Web site, which can then potentially be seen by fifty-seven million people for less than the cost of a poster on a billboard. Customers can find out about goods, order, and pay for them without going near a physical outlet. (Some goods, of course—flowers, groceries, computers—still need to be stored and delivered, requiring a physical supply chain.) Second, for some products (computer games, newspapers, software), the Internet reduces the cost of distribution. These products can not only be ordered but also delivered electronically, presenting wider scope, not just for marketing, but for enormous economies of distribution. On-line delivery thus offers an extremely inexpensive way to test-market new products, such as software and games. The main costs of such products lie in creating them in the first place: the costs of publicity and distribution can be insignificant, as can the costs of "manufacturing" extra units.

Even when the product is physical and requires physical distribution, new models are emerging with lower entry costs. A prime example is Amazon.com, a Seattle-based bookshop that displays on its Web site a catalogue of most of the English-language books in print. Apart from four hundred or so bestsellers, the company keeps no books in stock. Instead it passes on customers' orders to the nearby warehouse of Ingram Book, one of America's biggest booksellers, to be wrapped and sent out. The danger for Amazon is that its low-cost model is easily copied. The firm's future will depend on the skill with which it builds loyalty and a sense of "belonging" among its customers.

The combination of low entry costs, low marketing costs, and low manufacturing costs helps to account for the extraordinary pace of innovation in electronic commerce. Success may not be any easier to achieve in the electronic than in the nonelectronic world, but failure costs far less.

Better customer information

Better communications improves customer information. This happens in several ways. The costs of making information available, especially in catalogue form, decline; electronic mail allows businesses to alert customers quickly about new products or product changes; and customers can share some of the product tracking information available within a company.

Electronic delivery dramatically reduces the costs of distributing information. Not only does producing a paper catalogue cost substantially more than does producing one on-line, but copies of a paper catalogue carry costs for paper and postage, while an electronic catalogue costs the same, whether it is seen by five people or 500,000. In addition, customers ordering on-line type in their own order details, so their orders are more likely to be accurate—and less likely to need expensive retyping.[16]

Small, specialized companies benefit particularly from this economy, which enables them to sell to global niche markets. Aussie Lures, for example, runs a tiny business from a garage in Sydney, selling fishing lures via the Internet to customers around the world.[17] HotHotHot sells hot sauces and chili mixes through a mail-order catalogue and a store in Pasadena, California. The company's Web site, established in 1994, has come to account for 20 percent of sales, or twice the share of the paper catalogue. But, more important, while marketing costs eat up 22 percent of catalogue revenue, they reduce on-line sales revenue by only about 5 percent.

For companies with large product ranges, selling mainly on price and speedy delivery to other businesses, electronic information offers an inexpensive and quick way to update a catalogue. Quite apart from the cost savings, it offers easy searchability and a built-in order-taking mechanism. Fisher Scientific, for example, a Pittsburgh firm specializing in scientific instruments, puts out a 2,600-page paper catalogue that weighs nearly ten pounds, but still has room for only 45,000 of the 100,000 items that the company can supply. The company's comprehensive on-line catalogue constitutes one of the largest of its kind on the Web.[18]

The more frequently products are updated, the more useful becomes an inexpensive means of notifying customers about changes. *HotWired* and *Slate*, for example, two on-line publications, use electronic mail to tell readers about articles that might be of interest to them. Companies also increasingly use personal broadcasting techniques to keep customers up-to-date on product offerings. In December 1996, General Motors used personal broadcasting to send out an on-line notice about the launch of the 1997 Buick Regal to Internet users who had shown an interest in GM products. A click on the notice led users to a Web site where they could look at video clips of the new car.

Some companies have found ways to hook customers into their own internal tracking systems, giving them instant information on inventories and work in progress. Federal Express, the pioneer, created a Web site in 1995 that allowed customers to track the progress of their own packages around the world. (Before opening the site, Federal Express had sent customers special software on disks that would allow them to follow a package's progress.)[19] Other companies will increasingly make available to their customers non-commercially sensitive data already collected for internal use. Some airlines already allow people to follow the progress of an incoming flight, and some hotels allow customers to check directly on room availability. In the future, movie theaters may allow customers to check on seat availability. Retailers may eventually put on the Internet information about which of a chain's branches have certain products in stock, saving customers fruitless journeys or telephone calls.

More efficient markets

On-line technology creates two main opportunities to make markets work better. Where companies are willing to post the prices of products on-line, buyers can search easily for the lowest price. In addition, the Internet makes auctions easy and widens the number of bidders who can take part in them.

Information about products can be processed and presented on-line in many ways. A list of specific attributes can include maximum price.

That theoretically makes it easier for customers to compare prices and products and to search for the best deals. But companies will not always be willing to cooperate by putting their prices on-line. An experiment by Andersen Consulting to compare CD prices using software called "intelligent agents" ran into trouble when the biggest stores blocked it.[20]

Price comparison may be more commonly used by corporate traders, especially for standard items that are readily commoditized, than by individuals. If so, that will have important repercussions. Given that plenty of companies owe their profitability to imperfect information about prices, on-line trading will certainly squeeze margins. Having customers who know as much as they do about pricing frightens many companies. In 1996, Michael Fuchs, president of Germany's wholesale and foreign trade association, blamed the Internet (probably prematurely) for a loss of foreign markets for German goods. He argued that German companies offering to supply goods abroad at a given price would once have been fairly sure of securing an order; but now, after surfing the Internet, potential customers frequently quoted more competitive prices from rival suppliers. The problem would spread to retailers, he forecast, when a shopper with a credit card and a computer could order goods from around the world.[21]

While easy price comparisons will create downward pressure on prices, auctions may allow people to sell products for higher prices than would otherwise be the case. Auctions have proliferated on the Internet, to take advantage of its enormous reach. More than 150 sites now exist on the Web, such as AuctionWeb, which offers an auction room for people selling everything from rare Barbie dolls to barbecue grills and which conducted 330,000 auctions on-line in the first quarter of 1997. The largest on-line auctioneer, OnSale, sells $6 million worth of computer equipment and electronic goods a month.[22]

Several airlines also use on-line auctions to sell seats that would otherwise be empty, especially on weekend flights. Early instances were three auctions held in 1996, by Cathay Pacific Airways, which offered a total of 537 seats, mainly on flights from New York or Los Angeles to Hong Kong. In the latest, the airline offered 387 seats in various classes for travel in 1997 and got almost 15,000 bids, mainly from people who had not flown Cathay before. Cathay sees the auctions as a low-cost way to fill a few spare seats and as part of its main Internet strategy of building an electronic-mail database.[23] If airlines, why not other service industries that need to sell excess capacity, such as theaters and restaurants and tourist resorts?

Business traders have also been early users of the Internet's ability to create a global auction floor. The Chicago Board of Trade, for example, launched an Internet market in recycled materials in April 1996. The Internet provided an advantage here because the market in recycled materials encompasses a huge number of different products—several kinds of plastic, rubber, glass, paper, and so on—with highly volatile prices and a potentially large number of international customers. Transactions do not necessarily take place on the Internet, but traders can electronically mail bids to potential partners. In early 1997, the market was flourishing, with 185 registered users trading in ninety-six kinds of material.

In commerce of all kinds, customers do not necessarily always want the least expensive goods on the market. Quality, proximity, servicing, and delivery dates are often just as important (and reflected in prices). On-line trading will allow well-informed buyers and sellers to deal in all sorts of goods and services at prices that more accurately reflect supply and demand around the world. The market for rare Barbie dolls will thus become more like the market for stocks and shares.

Pressure on agents and intermediaries

The companies that have gained most by relying on imperfect information have been intermediaries, who have been able to earn a living in all sorts of businesses by putting customers in touch with services and then collecting a fee. What is their future?

Eventually every individual may have access to a screenful of the sort of price information that is now available to a broker or travel agent. Services such as buying real estate, equities, or theater tickets; renting holiday accommodation; or hunting for the best insurance quotation, will shift partly to an electronic marketplace. Sellers will simply display details; buyers will search through what is available.

Will that mean no future role for agents? Some analysts, notably Microsoft's Bill Gates, have argued that new technologies will wipe out intermediaries by creating a "friction-free capitalism." The Internet, he argued, "will extend the electronic marketplace and make it the ultimate go-between, the universal middleman."[24] Through the Internet, consumers will have direct access to information on all sorts of services. Retailers, travel agents, real estate agents, insurance brokers, and even dating agencies have no future in their present form. People will more often go directly to producers—whether it be goods, holidays, houses, insurance policies, or spouses. That transformation has already started: for example, 15 percent of packaged holidays in Britain are sold through direct-response advertisements on Teletext, an on-line information service displayed on television sets.

To survive, intermediaries will have to play a different role. They will benefit, just like other businesses, from the falling costs of setting up in business. Indeed, new sorts of intermediaries will emerge to take advantage of the new way companies will be organized in the future. The need for agents will be greater than ever, as information proliferates. By mid-1996, there were already, on one count, five thousand travel-related sites on the Internet. Even with refined search tools, the do-it-yourself traveler still faces a big expenditure of time and energy.[25]

Where information on prices becomes readily available, and products are in danger of becoming commodities, for which companies all charge the same, sellers will compete in other ways, such as convenience, comprehensiveness, or quality of service. More companies will follow the example of airlines, which compete on the basis of flight times, the quality of their meals, or the range of their routes. The complexity of such competition will ensure that many travelers still want specialist advice. British Airways has an efficient direct-sales operation—but 85 percent of its seats still sell through travel agents.[26]

If producers by-pass intermediaries—if, for instance, hotels increasingly offer direct booking systems or airlines sell a larger share of their tickets directly—companies will be taking on functions that they had previously outsourced. That will not be a good idea unless hotels and airlines are as good at selling reservations as at providing beds and flights, which seems unlikely.

In the future, individual passengers may find it easier to locate the cheapest air tickets for themselves, but traditional travel agents may respond by putting more effort into combining flights and hotels into holiday packages or by knowing more about the relative quality of service. Intermediaries will still be in demand, as long as they realize that their key to success is not the ability to have access to information but the skill to interpret it and market the result. Thus, in many industries, on-line markets are likely to coexist with agencies, altering rather than replacing them. Somebody buying a vacation might well want merely the cheapest holiday in a certain location. But others might also want a conversation with a human being about the quality of the various hotels at their destination. On-line services will encourage agencies to provide added value from human contact, such as advice, negotiating skills, and charm.

In addition, intermediaries often stand as guarantor between buyers and sellers. In the end, agents may run such markets, operating them like a real-estate agent's bulletin and taking a percentage of each sale in return for making sure that both buyers and sellers deliver what each promises. Intermediaries might thus become less expensive and more efficient—but they would not disappear.

Lower distribution costs

Products delivered on-line benefit from reduced distribution costs. Among the many products that are likely to be routinely available on-line in the future, some have been slow to get going. For example, although innumerable Web sites and chat sessions on the Internet are about music, the big record companies, which control two-thirds of the global music market, have been nervous about launching on-line music sales. They fear piracy and the impact on the retailers, who will long remain their principal distribution channel. In summer 1997, a number of trials were under way or planned for digital jukeboxes that would allow people to download music delivered to their homes by high-capacity links. Among the on-line markets that have been developing faster are videos, and airline tickets, as ticketless travel becomes the norm. But the most striking examples are financial services, games, information, sex, and gambling.

Financial services
Selling financial services electronically has several advantages. Long familiarity with automatic cash dispensers means that many customers

already have a great deal of experience in using a keypad and screen to check their bank balances and make other transactions. Financial services are already widely sold by telephone; indeed, the telephone will probably be the most important electronic means for delivering financial services to homes until well into the first decade of the next century.

In Britain, for example, Direct Line, a telephone-based insurer, grew from nothing to dominance in the car-insurance market in just a few years. Sharelink, a telephone stockbroker that has since been acquired by Charles Schwab, an American brokerage company, also had a big initial impact on its industry.

Financial services need interactivity more than do most other commodities. Buying a case of wine on-line involves merely scanning the details of what is available; the process will always remain more satisfying when it is possible to taste first. No such arguments apply to a customer buying stocks or making a payment. Instead, the financial information available on the Internet narrows the gap between amateur investor and professional broker.

So far, most financial-service companies have used the Internet mainly to provide information. From early on, Fidelity Investments, one of America's largest mutual-fund managers, has had a site that makes available a program with which parents can calculate how much to invest in a given plan to provide for a child's college education.[27]

Widespread on-line banking may still be some way off for any but the simplest transactions. After the chief technology officer at one American consulting company tried PC banking and gave up in frustration, he concluded, "Many of the minor problems that can be dealt with by a human when the customer is face-to-face or on the other end of a telephone turn into major problems when that same customer is communicating via a modem to a computer program."[28] Given the concern of many banks to cut staffing costs by keeping individual customers out of their branches, they have a big incentive to cure such problems.

On-line share-dealing has grown rapidly, thanks to ventures such as e.Schwab, the on-line subsidiary of Charles Schwab. In March 1997, about 700,000 of that company's accounts (one-sixth of the total) had conducted at least one on-line trade in the previous year. Forrester Research believes that assets worth $111 billion are already managed on-line, and that the total will grow to more than $470 billion by 2000, with perhaps ten million investors.[29]

This is still small stuff, considering that an estimated fifty-one million people in the United States own shares. In other countries, where private share ownership is much smaller, this area is likely to grow more slowly. On the other hand, on-line share trading may have as dramatic an impact on brokering costs as the initial rise of discount brokers had on Wall Street, where discounts allowed small investors to make trades without paying for unwanted advice. By spring 1997, at least thirty other American discount brokers already offered margins that matched or undercut e.Schwab.[30] And the low cost of the Internet share trading has encouraged some traders to offer the service free to customers with healthy surpluses on their accounts. In other words, share trading is becoming a loss leader for other services that are harder to commoditize.

Electronic games
Given adequate transmission capacity, on-line distribution is possible for electronic games. Increasingly, video games are being sold electronically: either downloaded from, for example, a cable-television link or played interactively over a network. Sega, one of the giants of the games business, has its own games channel; Nintendo is considering running games over telephone lines owned by GTE, one of the big U.S. local telephone companies.

Several companies, including Sony and Sega, are developing technologies for playing games on television screens connected to the Internet. The next stage is three-dimensional games in which players compete against other players, sitting miles apart in their own homes, but linked on the Internet or on faster, less congested proprietary networks. Some analysts expect "multiplayer on-line gaming" to be earning revenues of $1 billion by 2001—although much of that may come from advertising rather than subscription.[31]

Electronic publishing
On-line delivery makes obvious sense for publishers, for several reasons. First, it eliminates paper, the cost of which can easily be half (or more) of a publication's total costs. Second, on-line databases can be easily and inexpensively updated. New information can be put on a computer at a small fraction of the cost of printing a new airline

timetable or telephone directory, say, or revising a medical textbook. Third, information can be easily searched; publishers no longer need to decide whether all holiday accommodation in Maine or Morocco should be listed by town, by price range, or by size. Fourth, electronic publications can be embellished with sound or moving images. And, fifth, an electronic publication can be "customized" to meet a particular individual's requirements.

For specialist and academic journals, on-line distribution has become the obvious choice. Since some are distributed for the good of the academic community rather than in order to make a profit, cost savings become important. Other electronic publications have found it hard to sell subscriptions on-line.

By early 1997, about seven hundred newspapers, most of them American, had sites on the Web. But the *Wall Street Journal* was the first newspaper to charge subscribers for access to most parts of its on-line daily: it has acquired some seventy thousand on-line paying readers. Many other newspapers charge for special services: the Web site of the *San Jose Mercury News*, owned by Knight-Ridder, charges for access to back numbers and to the bulk of its news service, and the *New York Times* charges overseas readers. Some electronic newspapers considered charging and then backed off, for example, *Slate*, an electronic magazine bankrolled by Microsoft. Instead, advertisers support most electronic newspapers and "Webzines," paying to have hypertext links to their Web sites embedded in on-line journals.[32] Printed newspapers cost so little that the electronic versions must offer many additional services before subscribers will pay for them. Travelers abroad will enjoy a distance-destroying on-screen version of their hometown paper each day. Other people may prefer to buy an inexpensive bundle of paper, as long as it can be delivered to the door.

Sex

The sex industry (or "adult entertainment," as its promoters primly call it) has helped to pay for and pioneer lots of techniques of electronic commerce. The cinema, the videocassette (75 percent of films in the late 1970s were adult films), France's Minitel (an early on-line information network that carries what the French call "messages roses"), premium-rate telephone services, the CD-ROM: all were used, in their early years, partly or mainly to carry erotic content. Now, through the telephone and the Internet, the porn purchaser can choose from a global market of infinite variety.

Thanks to the telephone, a world market in pornographic chat flourishes. Punters phone a number in a foreign country that has relaxed rules about such things. (Countries with few local telephone numbers are especially prized, as the short number fools people into thinking they are making a local call.) Thanks to the accounting rate system, described in Chapter 2, revenue for the international call goes partly to the country in which the call is taken. The result is a $2 billion industry that generates more than 900 million minutes of telephone calls.[33]

On the Internet, porn purveyors are developing ways to personalize services, to allow voyeurs to "video conference" with the partner of their choice, and to check credit cards in an industry prone to fraudsters from all over the world. For the customers, the advantages are privacy, versatility, and a global selection. For electronic retailers, the on-line delivery of pornography is probably the best available guide to the future.

Gambling

Initially, much global electronic commerce may result from customers wanting to buy products that are legal in some countries but banned or hard to get in their own, such as gambling. An industry with revenues of $40 billion per year in the United States alone,[34] gambling came on-line early—partly as a way to go offshore and evade state controls.[35] The exclusive sports rights of digital television combined with the global reach, computing power, and immediacy of the Internet yields a technology that could have been purpose-built for gambling.

Several entrepreneurs offer virtual betting from the Caribbean or land owned by Native Americans in the United States. Some act as on-line bookies, taking bets from gamblers who wire their money to accounts in the Caribbean and place their bets with a few clicks of a mouse. Another betting service, prudently based in Belize, acts as a sort of gaming exchange, matching sports gamblers with one another. Governments will have to decide whether to legalize on-line gambling or to turn a blind eye.[36] Many such dilemmas will arise, in industries from brokering to pharmaceuticals, as electronic commerce becomes global.

Global reach

A company's market is no longer its locality or even its nation. Thanks to global toll-free numbers and the Internet's global reach, electronic commerce offers an inexpensive way to sell to the world. International mail-order and electronic distribution will be big businesses, bringing customers an unprecedented range of choice and driving price competition at the level of the electronic high street. It will create global brands and global pricing strategies—and global regulatory headaches.

Of course, snags are to be expected, especially for products requiring delivery. They include customs controls; the need for local distribution and (equally important for some products) servicing; and differences of culture, language, and standards (for voltages, television systems, sizing, and so on). These may slow the growth of global electronic commerce, but they will not prevent it. In time, global commerce will put pressure on national price structures and purchase-tax regimes. Once customers realize that companies sell the same product in different countries at varying prices, electronic communications may really start to create a global marketplace.

For companies trading in Europe, the development of on-line trade poses a particular dilemma. By 1999, several European countries will have started to adopt a single common currency, to be known as the "euro." Companies have habitually priced products differently in Europe's many markets, making nonsense of the region's attempts to develop a single, harmonized trading block emulating the single market of the United States. The advent of the euro has already forced companies to consider uniform pricing throughout those European countries that will adopt it. On-line trading may well accelerate the process, by making it simpler for customers to compare prices around Europe.

For companies, this global marketplace will provide an incentive to try to build global brands. Some products, like some film stars or sports champions, will use the electronic market to become more widely known and more dominant. This will bring a greater global uniformity: just as people everywhere know the Coca-Cola and Levi brands, so they may one day all name the same kind of coffee or camera or shoes. But there will also be scope for greater diversity, as millions of tiny companies find that they can cater to local tastes or niche markets more precisely than before.

. . . . .

For regulators, worrying about consumer protection, global advertising and distribution will bring special problems. People will now have ways to evade rules intended to protect them from fraud by unscrupulous investment scams or from taking drugs that have not been sufficiently tested. Local companies, still covered by such rules, will complain if regulators do nothing. But increasingly, buyers will have to protect themselves. In the future, signing up to a national consumer-protection scheme may become a way of adding value to a product. Reputable companies will see consumer protection as voluntary but desirable. And regulators will compete to devise schemes that combine adequate protection for consumers at reasonable cost and without unduly restricting the consumer's choice. Good consumer protection may thus become a tool of competitive advantage.

Methods of payment

Just as reliable transport and secure money are crucial to commerce in the physical world, so security and electronic payments systems will be at the heart of electronic commerce. Both payments and the secure delivery of electronically distributed products, such as music and information, require encryption strong enough to deter most hackers. Payments mechanisms have to solve two main problems: security and micropayments.

In the Internet's early days, one big security worry was the use of credit cards in electronic transactions: one 1995 survey found that, while 43 percent of respondents would happily write their credit-card number on a mail-order catalog form, only 5 percent would send it over the Internet.[37]

But they worried groundlessly: "I don't know of a dime that's been lost over the Internet," says Jim Barksdale, boss of Netscape.[38] In the future, thanks to the development of powerful encryption techniques, it will be far more secure to send a credit-card number over the Internet than to dictate it over the telephone or send it through the mail. In February 1996, two rival consortiums, one led by Visa and Microsoft, the other by MasterCard and Netscape, combined to develop a single proprietary software for safe credit-card use, called Secure Electronic Transactions. Testing began in 1997. As a result, credit cards seem likely to remain the principal way to make payments electronically. They already account for a sizable number of payments. Amazon.com, for example, selling books on-line, took in more than $3 million in revenue per month via credit cards in early 1997.

Credit cards are at present an unwieldy way to take micropayments. Many items for sale on-line will cost less than ten dollars—a game, for instance, or a movie—or even less than one dollar—a news clipping, a movie review, or a stock market quote. Just as nobody would try to pay for a daily newspaper with a credit card (and newspaper sellers would be less than delighted to accept one in payment), so some branches of on-line commerce ideally need a payment mechanism suited to micropayments.

To provide it, several companies such as First Virtual and CyberCash have developed electronic "money." Thus CyberCash has software that allows people to pay between twenty-five cents and ten dollars. The software is free to buyers, but the seller of a product pays CyberCash between eight and thirty cents per transaction. Such schemes have been only modestly successful. More probably, credit-card companies will fill the gap, aggregating a shopper's tiny payments and sending out a single bill at the end of the month.

Corporate Structure

· · · · ·

In the next few years, companies will define their relations with suppliers, distributors, and customers through the extent to which they use new communications links to make their internal information accessible to others. A customer will be able to see at a click what point in the production line a wanted product has reached; a supplier will be able to track directly a customer's stocks of a particular product; a distributor will feed invoices directly into a company's financial system. Companies will have to make difficult decisions about how far they will allow other firms into such intimate connections. A piece of information entered into a corporate computer will potentially be available to anybody with whom the company wants to share it. One of the most important questions for managers will be deciding where to draw the line.

The Internet is transforming the way corporate computer networks operate. Companies increasingly use computer networks to connect their branches or offices, to monitor sales, and to replenish stocks. Supermarkets and other stores collect information electronically at the point of sale and feed it into central data banks.

In the past, most of these computer networks have been proprietary: they needed special software designed to allow certain computers to talk to one another about certain things, but not to other computers and not about other matters. This made computer systems rigid and hard to adapt. When two companies merged, squads of highly paid specialists had to be retained to amalgamate their separate computer systems. When the business of a company changed, the computer system had to be overhauled to fit it for new tasks. Companies found their business activities circumscribed by their computer systems. Different types of computer software, incompatible with the software on computers in other parts of the office, proliferated. New company employees had to be initiated into the mysteries of the system. Staff in some parts of the company might not be able to communicate with staff in other parts if their computers were not attached to the same network. Huge expense and frustration resulted. Not surprisingly, vast investments in software and computers produced only tiny improvements in measured productivity.

The Internet is helping to change all this, thanks to its ability to link together almost all types of computers and software in a user-friendly way. All those isolated islands of information technology can at last be connected, theoretically at least. Businesses have overtaken the academic world to become the main users of the Internet. Most use it as a basis for intranets with the same software and network equipment as the Internet and speaking the same computer language. Intranets run on private networks within companies and among their branch offices, fenced off from the global Internet outside by firewalls that allow employees to look out but prevent others from looking in.

Multinational companies such as Levi Strauss, Eli Lilly, Lockheed Martin, and AT&T have connected thousands of their employees through intranets. So have many smaller firms. The benefits are immense. Staff using Web browsers can reach corporate Web pages or use hypertext links to flick through bulky corporate documents, such as contracts or safety manuals. Instead of photocopying a report for distribution, a company can simply put a single copy on the intranet for everybody to read. Bulky documents such as internal telephone directories and health and safety manuals can be distributed speedily and at little cost.[39]

Such systems bring problems, too, notably security. A company that happily puts its internal telephone directory on an intranet might think twice before putting financial information on the system, partly because of the risk that intruders might break through the firewall.

But because of intranets, the enormous investments that companies make in computer software and hardware are at last showing signs of paying off. Many companies have built intranets using their existing computer hardware and spending only a modest amount on software.

By using the Internet as a template, companies find they can experiment inexpensively, without buying new hardware or having to rejig a whole system. They can also connect disparate computer systems at little cost, linking up mainframes and personal computers and all the ragbag of different systems that many companies have accreted over the years. Instead of spending millions of dollars, as they might have done for such a task a decade ago, companies find that they achieve results with tens of thousands or even thousands of dollars instead. Companies are likely, as a result, to experience a revolution in the way they think about their internal networks. Rigidity will be replaced with flexibility; inaccessibility will give way to openness; fear of instant obsolescence, which has discouraged so much past investment, will dwindle. These changes will have far-reaching impact on employees; the location of company activities; suppliers; customers; and the structure of firms.

Employees

The intranet offers companies the opportunity to break down walls around as well as within them. Employees in different parts of the country—or the world—can keep in touch more easily with one another and with the head office. Intranets build on the shared information within a company that is one of its most important resources. Some companies deliberately build an intranet to exploit that resource—as in the case of Booz Allen & Hamilton, a management consultancy, which has set out to capture the knowledge of its partners, its main stock-in-trade.[40]

More intractable issues remain, however, including the questions of which employees should be allowed to see particular information, which employees will have the right to alter information, and, assuming some information is kept from some employees, how it can be done without making those employees feel hurt or excluded.

The Internet will change management culture. "I can hardly remember life without e-mail. It is woven into the fabric of how we operate," says Lew Platt, boss of Hewlett-Packard.[41] It would be difficult to find a chief executive of a large European or Asian company who felt that way. In many parts of the world, senior managers do not know how to type; some Europeans still ask their secretaries to print their e-mail and type their replies. That will change. (Once the quantity of e-mail builds up, the secretary's job will more likely be to filter the stuff, preventing overload.)

When the Internet becomes embedded in corporate management culture, it accelerates business decisions. Divisions between departments weaken when information that was once the private property of one department becomes available at the click of a mouse to every other. Hierarchies go. Junior staff can send messages directly to senior. Young employees (to the dismay of many Japanese middle managers) gain an edge over their elders. The Internet is a powerful force for democracy, even in the workplace.

Location of activities

Intranets allow a company with hundreds of scattered branches to use a single database. That is not just an economy: it also allows the company to locate the database wherever the skills to maintain it are most readily available. In all sorts of ways, better communications can change the dynamics of location.

Employees in different countries or regions, for example, can collaborate on the same project. A number of software companies have developed ways to make virtual reality available on the Internet, allowing the transmission of complicated computer-generated, three-dimensional images. Companies in the engineering business, especially the automobile and civil engineering industries, have been exploring ways to use virtual reality on intranets. In 1996, Division, a British software company, used the Internet to link five designers in three locations to work together on creating a Formula One racing car.[42]

Employees scattered over a wide geographic region can report to a few centralized managers. Frito-Lay, for example, has a roving sales force of ten thousand people who use handheld computers to record sales of more than two hundred grocery products, supervised by a mere thirty district managers.[43]

Relations with suppliers

Beyond the internal uses of intranets there lie even more intriguing possibilities. How much of their internal information should companies make available to big customers, key suppliers, or important distributors? All of these, in theory, can be connected to a company's intranet, allowing them to cooperate in new ways.

One model for the future may be the GE Trading Process Network, a secure Internet-based electronic network launched by the General Electric company in January 1996.[44] GE provides free software to current and potential suppliers, who can then register to be notified of GE product requirements, to receive electronic information, and to submit bids. The software manages the bids as they come back, eliminating those too low to qualify, handling subsequent rounds, and finally notifying the successful bidders of the result. Participating companies no longer need to telephone GE to ask for a faxed drawing of the parts the company requires. Instead, bidders simply download the relevant information from the Web site. They need nothing more elaborate or expensive than an ordinary Internet connection.

Through its Web site, GE now does $1 billion of business per year with about 1,400 of its suppliers, a sum that exceeds all consumer electronic commerce put together. The venture has had extraordinary effects. It has cut the the bidding process in GE's lighting division from twenty-one days to ten. Because requesting bids is simple, more firms bid and so the cost of goods has fallen by 5 to 20 percent. Before, it was complicated to include foreign suppliers in bids. Now, they account for 15 percent of orders. In the past, when machinery for one of GE's light-bulb factories broke down, the company would have invited bids to replace it from a few domestic suppliers. With the simplicity of the Web site, GE in the United States could now invite foreign bids as well and eventually awarded the contract to a Hungarian firm—at a savings of 20 percent.

But the impact of the Trading Process Network will eventually spread much further. Some suppliers are starting to put their own Web pages on the GE site, in the hope of attracting business from other companies. And other large companies with their own networks of suppliers are joining the TPN scheme. The first was Textron Automotive, a big car-parts dealer that spends $400 million per year on parts. It planned to join in summer of 1997, bringing another 700 suppliers to the network.[45]

Relations with customers

Location tracking will allow companies to build a new kind of relationship with customers, because they will be able to follow the use of their product beyond sale to the customer to the point of final disposal. Automobile manufacturers, for example, will be able to tell when their vehicles are reaching the end of their usefulness. That may signal the opportunity for a sale, or it may provide environmentalists with information with which they can ensure that manufacturers take responsibility for their products "from cradle to grave."

Already, manufacturers are using communications to build new relationships with their customers. Go into certain retail outlets of Levi Strauss, the blue-jeans manufacturer, and your measurements will be taken and whizzed over the Internet to the factory, where a pair of blue jeans tailored to fit will be produced. Levi Strauss thus builds a database that will not only allow customers to reorder (assuming that their shape does not change) at the click of a mouse, but will allow the company to mail them news of other products (in their size, of course).

Or consider a trend in educational publishing. Some American companies, such as McGraw-Hill, Addison-Wesley Longman, and Prentice-Hall, now enable professors to assemble a book tailor-made for a particular course, containing a chapter from one text they publish and a chapter from another, together with examples and diagrams from a third. A cigarette manufacturer has been exploring the possibility of printing cartons with the name of an individual smoker.[46] For tobacco companies, such data offers an alternative to cigarette advertising, which is increasingly regulated.

These examples represent a conversion of mass-market manufacturing into personalized service. Manufactured goods become increasingly made to order, a return to a world that vanished with the Industrial Revolution. Like television, products from shoes to newspapers will be increasingly tailored to a customer's individual quirks and needs.

The future of the firm

Will the communications revolution make the company an anachronism? The question is not a new one. In 1937, Ronald Coase, an economist who was later to win a Nobel Prize for his work, asked why workers were organized into firms instead of acting as independent buyers and sellers of goods and services at each stage of production. He concluded that firms were needed because they performed a useful role for their employees: they overcame lack of information and kept down transaction costs. It takes time or money or both to find out what products are being bought and sold; companies hold down those costs.

A newspaper, for example, may own its printing presses or may contract with outside printers. In Britain, the *Mirror* and the *Telegraph* groups own their presses; the *Financial Times* and the *Independent* do not. Printing capacity may be less expensive if bought under contract, but some of the savings will be offset by higher costs for coordination.

As communications improve, they reduce such transaction costs. Scouting the market for spare capacity and low prices becomes easier. Technologies such as electronic mail and computerized billing reduce the costs of dealing with arm's length suppliers. All kinds of services can be bought in. The result will be companies organized into smaller units.

Companies will come to look more like Hollywood studios. Between the two world wars, Hollywood stars tended to be employed on contract by studios. That had advantages: it stabilized the stars' incomes and guaranteed them work—but it also meant that the stars effectively subsidized the less good or popular performers.

In time, the studio system broke up. Making a movie today involves assembling a temporary "company": buying in the services of scriptwriters, costume-designers, technicians, producer, director, and cast. Each participant earns an income that reflects his or her individual worth in the market, rather than a pay scale imposed from above on a group of people of varying ability.

· · · · ·

Many companies, in the future, will look much more like a movie in the making. Their suppliers will work closely with them, tapping into their customer's electronic database to determine what they need and to supply it, just as smoothly as an in-house supplier would once have done. Some suppliers will have many customers, some only one.

If this scenario is right, one of the main impacts of the communications revolution may be to reduce the size of firms (although not necessarily of firms that own networks: see Chapter 6). This projection fits with a pattern that has already begun to emerge, strikingly in the United States: over the past two decades, the size of the average American firm has been shrinking. The more companies have invested in information technology, the smaller they have become. Whether measured by the number of employees, sales, or value added, companies are smaller now than they were in 1970: the average number of employees in American firms has dropped by a fifth.[47]

A knowledge-based company can farm out a much larger share of its activities than can a firm in a traditional heavy industry. Computers and communications make it possible to turn companies into networks of independent workers, specializing in what they do best and buying in everything else. So people will work in smaller units or on their own. Many may work from home, or from purpose-built condominiums of small offices. High-quality communications will make working from home more of a pleasure, especially for "knowledge" workers. Again, this will fit in with a pattern that has emerged, most clearly in the United States. Almost one American worker in five currently works from home for at least some of the time.[48]

The pattern has implications for finance, too. As the population of the rich countries ages, their savings will rise relative to their incomes. But those savings, instead of being channeled into a relatively small number of large companies, will ideally be spread across an immense number of small firms. Investing in small firms is frequently riskier than investing in large ones, simply because they tend to be more specialized and so more vulnerable to setbacks in their particular markets. Besides, as small firms will be essentially knowledge-based, investing in them will mean owning not physical assets but people. Investors will have to become good at judging the creativity and ingenuity of the people who will increasingly make up the main assets of most companies and at reassuring themselves that their investment is secure against these human assets walking out the door.

How will that investment take place? Remember that global communications will ensure that savings can surge around the world in search of the highest returns. New institutions will have to grow up, to funnel these savings into the small, fast-growing, risky businesses of the rich world: new intermediaries, skilled at putting together portfolios of small companies. Their specialization will not be producing goods, but bundling risks, in the way that large companies do now internally. They will become a new sort of mutual fund or investment trust.

Perhaps the most striking point about the changes facing companies is that they are driven not just by communications but by other forces as well. Communications merely happen to be reinforcing many of the trends that are already unfolding: the communications revolution is working with the grain of corporate change, rather than against it. That is why, ultimately, its influence will be so great.

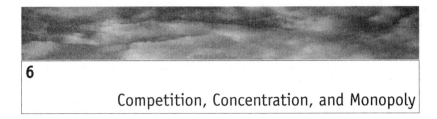

6
Competition, Concentration, and Monopoly

Competition in communications is new and enormously important. Fifteen years ago, people had to rub along with one national
monopoly telephone operator (or one local telephone monopoly, in the
case of the United States); a handful of big broadcasters or even a single one, run by the state; a solitary cable company. In few other industries in the capitalist world has this level of monopoly been allowed and
the spirit of competition so crushed.

But of course competition is just as important in communications as
anywhere else. Competition brings new players who have launched new
kinds of service: satellite television, mobile data services, wireless telephony, and the Internet itself. Competition alone guarantees that consumers will really benefit from the technology races now being run by
the telephone, television, and Internet. Only competition, among the
three and within each, has the power to drive down the premium prices
still charged for distance and to bring consumers the variety and choice
that technological change promises.

Competition, however, does not come easily in communications. The
opportunities to restrict access and to build monopolies have been greater in
communications than in many other businesses. Even once governments
pass laws to allow competition (such as the 1996 Telecommunications Act
in the United States, described by Reed Hundt, chairman of the FCC, as
"The biggest global effort to de-monopolize any industry sector that has
ever been mounted"[1]), months or years may go by before much changes.
Competition becomes bogged down in courtroom battles or simply fails to
deliver: in Britain, where competition began in 1984, BT, the former state
monopoly, still has 90 percent of the residential market and 74 percent of
the business market; in the United States, AT&T has 57 percent of the long-

distance market, thirteen years after it was split up; and Japan's NTT, theoretically competing, has a vice-like grip on local telephony.

Indeed, the immediate aftermath of deregulation has sometimes been more concentration, not less. In the United States, the Telecommunications Act of 1996 was succeeded by a merger between Bell Atlantic and Nynex, two large regional telephone companies; after Japan's liberalization proposals at the start of 1997, Japan Telecom, a domestic long-distance carrier, and International Telecom Japan, an international carrier, plighted their troth.

At the same time, new monopolies appear to be springing up beside the old ones: most obviously Microsoft with its grip on operating systems, but also Cisco, which dominates the market for Internet routers, Intel (microprocessors), Intuit (personal-finance software), and BSkyB, Britain's satellite-television broadcaster. Clearly, something about the communications business breeds concentration.

In fact, several of the industry's characteristics—such as the reliance on networks and the importance of standards and systems—naturally encourage monopoly. Up until now, many governments have been happy to go along with communications monopolies, especially where these have been owned for the most part by the governments themselves, which have then universally regulated them more intrusively than they have other industries. Now governments have to switch from being protectors of monopolies, often monopolies they themselves were running, to being champions of freedom.

In fact, signs of change can be seen. Why? Because governments realize that a competitive communications market is overwhelmingly in the national economic interest. While the opportunities to frustrate competition in communications remain enormous, so are the gains from making competition happen.[2] Indeed, that realization persuaded some seventy countries, many of which still assume that communications should be a state business, to agree at the World Trade Organization in February 1997 to open their markets to foreign competition. Change will come slowly, but change is on the way.

These trends raise several questions. Why do monopolies still have such a grip in these industries? Does it matter? And if it does, what should governments do about it?

By the end of the century few countries in the developed world will have a single monopoly telephone operator, a limited number of television channels, or serious restrictions on cable services. Ten years into the next century, communications will be much like any other industry: consumers will have won.

Why Communications Is Monopoly-Prone

. . . . .

Look at the dominance of Microsoft's key operating system for PCs or Netscape's Internet browser. Think of the importance of your telephone number and the costs and inconvenience of changing it. Think about how a program guide directs the eye to certain television channels or how a search program on the Internet selects sites. These are gateways: control them, and a company can lucratively steer customers to particular products and services. Broadly speaking, such gateways can be built in three ways: by controlling the hardware of communications; by building the dominant software; or by developing or buying the "must have" content.

Technological change has widened hardware gateways in the past few years, but they still exist. Messages can be sent to or from a home or office in only a limited number of ways, for example, and they are not yet interchangeable. Satellites have a finite number of transponders from which television programs can be beamed, and the wiring of the last mile to a customer's home typically has less capacity than the rest of the system. Internet routers can handle only so much traffic.

Technological change has not reduced the number of software gateways between customers and the services they want to buy. Instead, it has created new opportunities for market power. For example, users now require encryption systems. Software for encrypting and decoding cable or satellite television is already an obvious gateway: in Germany in 1996, two different alliances of digital satellite broadcasters fought to establish their systems. Operating systems constitute another software gateway. The obvious example is Microsoft: more than 80 percent of the world's personal computers contain Microsoft's MS-DOS or Windows operating systems; and Microsoft is by far the biggest supplier of desktop software applications. Navigational tools, too, are increasingly essential. Browsers that help Web users to leap from one page of information to another and the electronic program guide that helps viewers pick television programs are important gateways: control them, and you influence what people look at on-screen. Subscription management and customer relations software will help companies run the electronic shopping malls of the future, controlling the link between retailers and their customers and gathering large amounts of commercially valuable information about the behavior of both.

Control over desirable content is the ultimate source of market power: think of the value to be squeezed from owning a useful database or the rights to some immensely popular sporting event. You may swap your cable operator for satellite television or your PC for an Apple Mac, but you will not regard a minor-league game as an acceptable substitute for the big match. The earning power of celebrities reflects the strength of this monopoly. The dozen or so movie stars whose names are known worldwide comprise one of the world's scarcest resources. Lots of people can act, but there is only one Kevin Costner.

Those are the gateways. But what is it about communications that produces so many of them? Both high-tech industries and communications have several characteristics that make them prone to concentration. Big is not necessarily Bad: only when Big is used to fleece customers do antitrust regulators usually see a case for moving in. But, when market entry is difficult, as is often the case in communications, the forces of competition will be weaker and market concentration greater.

The industry, in part, inherited its current level of concentration in communications: all countries have had a long period with only a single giant telephone company. (The United States is unusual in having split up AT&T in 1984, but it left the local monopolies intact.) Most have had one giant broadcaster. (Again, the United States had three networks; Japan with NHK and Britain with the BBC were more typical in having one main broadcaster for many years.)

But new giants have emerged in the newer communications and high-tech industries. Even in the older industries, the enormous number of strategic alliances and mergers—Nynex and Bell Atlantic; BT and MCI; News Corp and MCI; Sprint, France Telecom, and Deutsche Telekom; Disney and Capital Cities/ABC; Kirch, Nethold and Canal+; AT&T and McCaw; and so on—suggests that the level of concentration in the industry at best is static but may be rising.

In 1996, concentration began to affect the Internet, as well. The myriad small access providers started to lose ground to large telephone companies, such as AT&T, which began to offer inexpensive access on a huge scale, using their market power to tempt their existing telephone customers and their strong brand to reassure the technophobic. At the same time, Microsoft began an assault on Netscape's dominant position as a software gateway to the Internet. By giving away thirty-eight million copies of its Netscape Navigator, a browser that steers people through the complexities of the Internet, Netscape in a mere two years acquired 85 percent of the market. But then, Microsoft launched its browser software, and because it was incorporated in most new PCs, along with Microsoft's operating system, it rapidly began to gain market share.

Three characteristics, some unique to communications and others common to other industries, underlie the tendency of high-tech and communications markets to succumb to concentration: the economics of networks, the predominance of systems, and the importance of standards.

The economics of networks

Think of water, electricity, railways, airlines, and roads. All are networks. All share one characteristic: the bigger they are, the better they work. A road that runs merely from your front gate to that of your neighbor is far less useful—to both of you—than one that joins you both to the national road network. Big networks also have economies that smaller ones lack: an electricity network that covers one small town needs more back-up capacity than a big network sharing power over a whole region or country.

These effects are obvious in the case of the telephone. To join a telephone network with half a dozen other subscribers would be pointless: the more people with whom a telephone network connects, the more useful it is. If mobile telephones could not link into national fixed-wire networks, they would be no more useful than old-fashioned "walkie-talkie" radios. So a large network, with millions of customers, has a huge advantage over a small one or one that is just starting up.

The advantages of large networks create a bias towards monopoly and explain why network businesses have frequently tended to be monopolies, and often state-owned monopolies at that. Many industries in and around the communications business are based on physical net-

· · · · ·

works, such as cable-television systems or computer networks. Sometimes the network is more "virtual," consisting of the users of a particular sort of software, for example. New networks have sprung up: the Internet, for example, or the decoding equipment needed by customers of digital satellite broadcasters.

Even entirely abstract networks can still be powerful. If half the population watches the Super Bowl or the World Cup Final, the other half pays a certain social cost the following morning, when it is excluded from water-cooler gossip. Perhaps the most powerful network of all is language: anybody unlucky enough to grow up speaking only Welsh or Estonian will be at a tremendous disadvantage compared with those of us fortunate enough to have English as our mother tongue. As the French and Germans are unhappily aware, a language that is the mother tongue of some 320 million people and understood by perhaps one-third of humanity[3] has an enormous advantage over all its rivals.

The prevalence of systems

The products of the communications industry are rarely used in isolation. Instead, several compatible components are designed to work together to form a "system." A computer, say, without an operating system; or an operating system without applications, such as word-processing programs or spreadsheets; or a television set without receiving equipment, such as a satellite dish and a set-top box to decode the signal; or even the wrong sort of telephone jack for a socket—all are useless.

Systems are so important that companies controlling one part frequently lever their way into others. Microsoft's behavior constitutes a classic example of the use of systems to extend market dominance. By "bundling" or incorporating software programs into its operating systems, Microsoft has gobbled up much of the initial market for stand-alone software. Buy a personal computer with Microsoft software and you will not, for example, need to buy a separate fax program, as you would once have done. Why, then, should anybody pay extra for stand-alone fax software?

Communications systems promote vertical integration in the industry. Britain's BSkyB satellite broadcaster, for example, controls a large share of the transponder slots on the Astra satellites that beam television programs into Britain; BSkyB is Britain's main producer of programming for pay-TV; it has the most important football rights; and News DataCom, which makes encryption software for conditional-access television systems, is linked to News International, the largest shareholder in BSkyB. In 1998, BSkyB plans to begin digital satellite broadcasting. It expects to subsidize the set-top boxes that customers will need to watch its programs, and its software for encrypting and decoding programs will be built into the boxes. As a result, any other broadcasters who want BSkyB's customers to watch their programs will almost certainly have to pay BSkyB for the privilege.

The importance of standards

Common standards are essential to networks and systems: they allow users of networks to communicate with one another, and they make each different part of a system compatible with all the others. They are the glue that holds communications together. A common standard makes an industry work efficiently: think of the irritation involved in taking a laptop computer abroad with a pocket-full of telephone jacks and a bag of different electric plugs, or the annoyance of having to convert a videocassette recorded by a friend in Germany to play on a television set in the United States, or the maddening way that a mobile telephone that works throughout Europe cannot be used in the United States. Compare that with the ease of using a telephone: dial a number anywhere in the world and it connects (usually), even though the call may have to pass across a dozen different telephone networks and finish on a machine in another part of the planet.

The Internet provides the ultimate example of the boon of a common standard: it allows lots of different computers (including PCs and Apple Macs) to talk to one another. Another essential standard—human languages—affect worldwide systems: air-traffic controllers use a standardized set of English words and phrases for their commands to encourage efficiency—and safety.

Standards bring big benefits. Think of the technology for receiving digital television. In the United States, a Grand Alliance of the consumer electronics, broadcasting, and computer industries agreed in

1996 on a common technical standard for transmitting digital television signals.[4] This agreement, which will allow consumers to buy wide-
screen sets capable of receiving high-definition digital television some
time in the course of 1998, has three benefits: it reassures consumers
who might otherwise have postponed a purchase until a single standard emerged; it allows manufacturers to enjoy the economies of one
large market; and it may set a standard that the rest of the world will
follow, thus bringing the same benefits of confidence and economy to
the global market.

But standards can also create problems. A widely used standard is
extremely hard to change. Standards that are owned by companies,
rather than public property, become licenses to print money. Changing an
established standard costs a great deal. Think of the QWERTY keyboard,
universal in the Anglo-Saxon world and supposedly designed to force
typists to hit the keys slowly to avoid jamming the old manual typewriters. (But see below.) The QWERTY standard has become so powerful that
it survived into the age not just of electric typewriters but of PCs. The
alternative Dvorak keyboard, designed to allow people to type more
quickly and accurately, has never taken off, primarily because millions of
machines exist with the "wrong" keyboard and millions of typists have
learned—and continue to learn—to use it. To change the keyboard would
involve huge expenses of money and time.

The QWERTY keyboard, like the English language or the Internet
protocol, is an open standard, not a proprietary one. No company
makes an extra penny or two each time a new QWERTY keyboard is
sold. Many other standards, however, are established by companies.
Microsoft's operating system is one of many examples. Writers of computer-games, for instance, have had to accept that Nintendo and Sega,
Japanese makers of computer-games machines, also make cartridges
and influence the content of games.[5]

In communications, the rapid evolution of new technologies creates lots of opportunities for trying to set an industry standard. In
Japan, two rival consortiums, one led by Toshiba and the other by
Sony, fought all through 1995 to set the standard format for the Digital Video Disk, a tiny device on which an entire film can be recorded;
in 1996, the Toshiba camp emerged victorious. Another battle rages
over the standard for digital mobile telephones: the GSM (Global System for Mobile Communications) standard, developed by European

companies, had been adopted in 110 countries by the spring of 1997; but Japan in early 1997 seemed likely to adopt CDMA (Code Division Multiple Access), one of two alternative standards developed in the United States.

Many people worry that a standard that emerges simply because one company gets there first may not be the best one for the market. Of the three main standards for analogue television—the United States's NTSC (said to stand for "Never Twice the Same Color"), France's SECAM, and variations of PAL ("Perfection At Last") used in most of the rest of Europe—the best is generally thought to be PAL. The United States, by getting into the market earliest, tied itself to an inferior standard that produces television pictures with much worse definition than those produced using the other two standards.

But not many industries trust government to set standards (understandably, given fiascoes such as the backing of an analogue version of high-definition television by the Japanese government). And in any case, the market will usually get it "right." One study argues that the VHS standard for videocassettes, regarded by cognoscenti as inferior to the defeated alternative Betamax, actually has important advantages: it allowed longer recording times, its price was lower, and its picture quality was not as inferior as some claim. The same study also argues that the Dvorak keyboard does not necessarily allow faster typing than does QWERTY.[6]

Does Concentration Matter?

· · · · ·

Given the natural tendency of the communications business toward concentration and monopoly, should governments simply sit back and do nothing? For many politicians, that has always been an attractive course. Governments, often owners of the telephone monopoly, have found it to be a handy source of revenue, and the absence of competition in broadcasting has allowed politicians to keep a hold on a powerful political medium. Competition is always disruptive and unpredictable: workers in big, established telephone companies and broadcasters fear for their jobs, and governments have responded to this powerful lobby. Finally, with competition, governments and reg-

. . . . .

ulators worry that a once politically compliant industry will become less responsive (not surprisingly: Europe's satellite-television channels meet few of the standards of political balance required of land-based broadcasters).

These arguments may weigh with politicians, but they do not make much economic sense: they do not demonstrate that most people will be better off with a high degree of concentration—or indeed, with a monopoly—in the communications industry. Three other arguments support this counter position. First, monopoly allows cross-subsidy. In the case of telephones, for instance, business customers help pay for telephones to isolated rural areas. Second, the global market demands that communications businesses be big in order to compete successfully. Third, and most important, concentration makes communications networks more efficient.

Cross-subsidy and universal service

Theodore N. Vail, the first chairman of AT&T, built that national giant on the back of the slogan, "One policy, one system, universal service."[7] The local telephone monopoly has traditionally had an obligation to provide universal service (although this has rarely been spelled out in its license). Now, whenever telephone companies face a loss of monopoly, they talk a great deal about universal service. Without the profits from monopoly, they say, they cannot guarantee that every part of a country, no matter how inaccessible, and every citizen will have access to affordable basic telephone service. So, when markets are opened to competition—as, for instance, in the United States through the passage of the 1996 Telecommunications Act—a decision must be made about who should pay the bill.

That sounds fair. But, in fact, it is often difficult to work out what universal service really costs. Further, it is not obvious what "universal service" means. Does it mean that all citizens should be charged the same telephone tariff? That all schools should have guaranteed Internet access? That everybody has access to network television? That people with disabilities or low incomes should have special access? Finally, why should some subscribers (city dwellers and businesses) pay extra to subsidize others (rural dwellers and residential customers)?

Once the obligations of "universal service" are clearly spelled out, they become harder to defend. True, the telephone is an essential part of modern life: it can be indispensable for finding a job, reaching a doctor, or staying in touch with family and friends. But exactly the same could be said of the automobile. Should governments provide an automobile for every citizen or fix vehicle prices?

Competition forces governments to look more closely at the cost of universal service and at better ways of providing it. A path-breaking study by Australia's Bureau of Transport and Communications Economics, published in 1988, asked operators whether, if no universal-service obligation existed, they would choose to provide it. The survey found that Australia's then telephone monopoly served between 110,000 and 220,000 uneconomic customers at a cost of between 1.5 percent and 3.1 percent of its annual revenues—far lower than the company's own estimate of more than 19 percent. Other studies, in other countries, have all shown similar results. The additional cost of offering a universal basic telephone service is small.

Indeed some new industry entrants say that universal service should be regarded as an opportunity rather than a burden. Certainly every country, rich or poor, that has allowed competition has seen telephone density— the number of lines per head—increase. Even in Britain, a mature market, more than 10 percent of the subscribers wooed by the cable companies have been people who previously did not have a telephone.

This roughly parallels what happened when the American airline industry was deregulated. Congress insisted on setting up a fund to subsidize two flights per day to 150 small cities that legislators feared would otherwise lose service because the big airlines had insisted that they could not make the routes pay. In fact, the fund was not needed: new commuter airlines discovered that they could serve these routes profitably using not the big expensive jets of the large airlines, but smaller turbo-prop craft.[8]

Concentration and global competition

After a decade of dithering about whether to break up NTT, as the United States broke up AT&T, the Japanese government finally decided at the end of 1996 to leave the giant intact—and, indeed, to allow it to

. . . . .

enter the international telephone business. Why? Mainly because of a proposed merger between Britain's BT and America's MCI. Suddenly, Japan could see that size might be essential in global communications.

Again, the airlines offer a vision of the future. Just as global competition has forced the world's airlines to form alliances and share codes on busy routes, so telephone companies are joining together to provide end-to-end service for customers. In both industries, these alliances worry regulators. A liaison between British Airways and American Airlines, announced in June 1996, gives the two companies control of 60 percent of the flights between Britain and the United States. A deal joining France Telecom and Deutsche Telekom with America's Sprint links two national telecommunications monopolies.

In these cases, what suits the carrier and a minority of big business customers may not be best for the majority of customers. The solution, discussed below, is to ensure that concentration does not become monopoly and to encourage competition.

Concentration and efficiency

The characteristics that predispose networks to concentration also make them more efficient and hence more useful to customers. Systems that fit together well work better than those that do not, and common standards save time and money.

The simple tendency of the communications industry to concentrate will not necessarily make consumers suffer. Indeed, they may benefit if the big company delivers efficiencies that smaller ones cannot. For many customers, for example, having AT&T or BT as an Internet provider may have benefits over a little company with fewer resources and less muscle to push down prices. Most people would rather cope with mastering a single computer operating system, even if it enriches Bill Gates, than half a dozen different ones.

Regulators must find ways of ensuring that communications industries enjoy enough competition to guarantee innovation and to squeeze prices, while consumers at the same time enjoy the benefits of efficient networks, systems, and standards. Market dominance is not wrong—as long as it can be made to coexist with flourishing competition and easy access for new entrants.

The benefits of competition

Competition is an especially powerful force for improving communications. In countries with competition, telephone ownership rises quickly, prices decline, and networks modernize briskly. Substantial evidence exists for these positive effects. First, competition on international calls encourages fast growth in traffic.[9] Competition also brings quick installation of new telephone lines. People in competitive markets make more calls: subscribers make around three thousand calls per line per year in the United States, for example, as compared with fewer than one thousand in Germany, where the telephone monopoly's advertising slogan for years was *"Fasse Dich kurz"* ("Keep it short").

New entrants bring better services. In Britain, the cable-television companies, which also provide a low-cost telephone service, have introduced free itemized billing and other schemes to encourage poorer households to acquire a telephone. New entrants generally install state-of-the-art equipment, encouraging established operators to do the same. They have, for example, encouraged rapid take-up of mobile telephones. Where countries licensed more than one mobile-telephone operator, ownership soared in the early 1990s; where they did not, it remained flat.[10]

The Internet also develops more rapidly in countries where telephone services compete than where they do not. A study by the OECD in 1996 found that Internet access in countries with competitive markets has grown six times faster than in monopoly markets.[11] Furthermore, Internet use is much cheaper where there is industry competition. In Switzerland, anybody using on-line services for an average of forty hours per month pays up to twenty times as much in call charges as a similar user in Canada.[12]

Not only rich countries benefit from competition. Less developed countries do so, too. In the Philippines, the telephone service was so bad in the early 1990s that when Lee Kuan Yew, prime minister of Singapore, visited the country, he joked that 98 percent of Filipinos were waiting for a line and the other 2 percent for a dial tone.[13] Then, in 1993, President Fidel Ramos signed an order obliging licensees of cellular and international franchises to provide fixed-link telephone service in under-served parts of the country, a regulation that will quadruple the country's telephone lines.[14] (See Figure 6-1.)

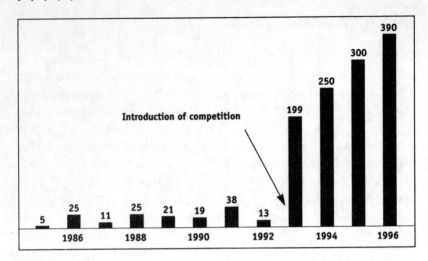

Figure 6-1   New Main Lines Added per Year in the Philippines, 1985–1996 (in thousands)
Source: International Telecommunication Union.

In television and radio, competition also multiplies choice. Americans are used to having many television and radio stations, but for most of the rest of the world, choice is a relative novelty. The benefits of choice in television and radio are more difficult to measure than the benefits of improved telecommunications services. But the speed with which the audiences of the established public broadcasters have declined—in, for instance, Germany and Spain—suggests that many viewers and listeners take advantage of their new station options.

As scope for new multimedia services increases, so will the benefits of competition, such as being able to try out new services. In uncompetitive markets, the barriers to entry and the costs of failure will be greater than in competitive ones. The vibrant experimentation on the Internet results from the negligible barriers to entry—ordinary people can try out new ideas as easily as can large companies.

The challenge for regulators, then, will be to ensure competition without destroying the genuine benefits that can come from market dominance in communications. That means opening markets; setting clear rules for competition; and, sometimes, giving new entrants a helping hand.

Creating Competition

. . . . .

Many people assume that, once markets are open, competition will automatically flourish. In fact, given the strong pressures for concentration in the communications business, that may not necessarily happen.

Opening markets

Without open markets, competition has no hope. In the mid-1990s, many steps toward opening markets were taken—on paper. For instance, the countries of the European Union (apart from the poorer ones, who chose to wait a bit) agreed to open every telecommunications market—for voice, mobile, and satellite—at the start of 1998. Some seventy countries agreed at the World Trade Organization to open their markets to foreign telecommunications competition and to set rules for fair global competition, starting in 1998. America's Telecommunications Act of 1996 allowed a free-for-all in communications: long-distance carriers entered local telephone markets, local carriers offered cable television, cable-system operators offered telephony, and so on.

But these admirable decisions constitute only a first, small step, because competition in communications can be of two main sorts: in services (for example, television channels or long-distance telephone services) and in infrastructure (for example, cable systems and telephone networks). Often, one company is in both businesses: telephone companies, for example, both own wires and offer services (Internet access, voice mail, international call-handling). In the television business, Britain's BSkyB is the country's largest producer of pay-television channels, as well as the only satellite broadcaster. The same holds true for France's Canal+. In the United States, TCI and Time Warner, the two largest owners of cable systems, both also own giant media groups.

The combination of service and infrastructure represents a problem for new entrants—for Rupert Murdoch, for example, trying to get his cable news channel onto Time Warner's New York cable system, say, or for the local telephone companies trying to compete with Japan's NTT. Building new infrastructure is expensive, and in many countries it has been discouraged or forbidden by law. But without it, trying to achieve competition becomes rather like trying to set up a new supermarket by using a rival's in-house distribution system. How much (to use the supermarket analogy) should a newcomer pay for distribution? Can the newcomer use its own drivers and paint its name on the delivery trucks? If so, who pays? And how are disputes settled? By the existing supermarket board? Or by some outside regulator?

In time, technology will provide an answer. Most people will have more than one way to receive television or telephone signals. The Internet has pointed the way. Once people can more or less interchangeably use different kinds of infrastructure (their television receiver, their computer, and a telephone line), competition will be easier to ensure. Consumers will be able to choose among several providers of services and infrastructure as readily as they now choose which supermarket to shop in. But that day, in most countries, remains a long way off. Until it arrives, governments need to set rules for competition.

Setting clear rules for competition

In opening communications markets, regulators must start by deciding what to do about existing giants. Two possible answers present themselves. The United States offers one model: force the dominant company to split up, as the United States split AT&T into long-distance and local components. Britain represents the other: allow the dominant company to remain intact, but regulate it "asymmetrically," that is, impose obligations on it that smaller rivals do not incur. That approach has been followed by Oftel, Britain's telephone regulator, and by Austel in Australia.

It is no accident that the two approaches have been discussed and refined mainly in terms of the telecommunications business rather than television and radio. Media regulation has focused mainly on content and ownership. Most countries have been far more concerned about sex and violence on television or about whether foreigners should be allowed to acquire national newspapers or newspaper owners to acquire television stations than with breaking up existing large concentrations of media power. Italy, whose media baron, Silvio Berlusconi, briefly became prime minister, is a rare exception.

The telecommunications giants start with lots of advantages. They may control the billing system, a powerful link to the customer. They have well-known brands. Their business already enjoys all the economies of a large network. Their fixed costs may be largely depreciated, and they will have a steady cash flow to invest in new services. Deutsche Telekom, for instance, used revenue from its monopoly services to halve the price of the services that were liberalized first, such as data transmission. The result, at least in the past, has been to confine competitors to a tiny share of these markets.[15] It thus becomes easy for an incumbent operator to fight off would-be rivals.

For regulators to deal fairly with this situation they must be independent, and they must establish rules governing both interconnection and numbering.

Independent regulators

Establishing competition becomes much tougher when, as in a large number of countries, many of them developing, the regulator is actually the big telephone company itself. Vigorous competition cannot be established under these circumstances. In other countries, such as Japan, where the regulator is a government ministry, political meddling results. An independent regulator, such as the U.S. Federal Communications Commission or Britain's Oftel, is undoubtedly the best answer.

Rules about interconnection

The nub of the problem for regulators is establishing the terms under which those who do not control the dominant network gain access to it. The interconnection issue comes up in many network industries and often determines whether new entrants live or die. In airlines, the equivalent issue is allocation of airport slots; in electricity, the terms on which independent power generators can sell power through the grid.

In the telephone system, the big monopoly—or, in the United States, the local monopolies—control access to the local loop. As almost every call either starts or finishes on the local network, the charges for local access set a floor to the price new entrants can offer. In the case of television, the equivalent debate centers on the terms under which companies with programming gain access to the delivery system, whether it is a cable network or a satellite service.

Big telephone companies sulkily refuse to tell their smaller rivals where the interconnection points on their network are, or they simply refuse to answer such inquiries. They insist that competitors who share the network should pay some part of its cost. Fair enough. But how big a part? Most of the costs of a telephone network are capital: the electrical impulses that new entrants send down the wires do not add to the wear and tear in the way that, say, a railway carriage might wear the track over which it travels. Moreover, in many countries, the capital cost of the telephone network has been stumped up by taxpayers. Why, new entrants often ask, should the newly privatized incumbent then seek to profit by charging for access to it?

Besides, a network owner allowed to recover costs—on some definition or other—has no incentive to reduce them. How, too, can the true costs of running a network be decided? In the United States, that dilemma persuaded the Department of Justice in 1984 to order AT&T to hive off its local operations from its long-distance business. But while the split created competition in long-distance traffic, it left a monopoly in the local network. So the arguments about interconnection continued, with AT&T, MCI, and Sprint, the three main long-distance carriers, grumbling nonstop about having to pay over to the Bells 40 to 45 percent of the cost of a call.

Now, thanks to the 1996 Telecommunications Act, the long-distance carriers have won the right to break into the Bells' markets. But the arguments over interconnection, far from ending, have been one of the trickiest aspects of implementing the 1996 Act. In New Zealand, too, disputes over the way to measure interconnection costs have made lawyers rich and customers cross.

Plenty of other interconnection quarrels are rumbling along. First, what should Internet service providers pay telephone companies for access to their network? At present, in the United States, "enhanced services" such as the Internet are exempt from local access charges. Second, what should countries pay for access to one another's telephone networks? Third, what should mobile-telephone companies pay for access to fixed-wire networks?

Interconnection issues will also crop up in television over the proprietary "conditional access" systems that will allow pay-television viewers to receive and decode a particular signal for digital television. In Germany, the necessary software has been designed to Kirch's specifications, in France to those of Canal+, and in Britain by a sister company of BSkyB. Because viewers are unlikely to buy more than one set-top box for decoding encrypted television signals, Europe's regulators have demanded that the boxes be designed so that they can receive rival digital-television services. But they will be able to do so only on the terms set by the dominant television companies—or imposed by regulators. Stand by for a rerun of all the arguments that U.S. cable-television systems operators have made over the years regarding the terms on which they carry their competitors' programming.

Stand by, too, for a whole new set of interconnection arguments, as the different parts of multimedia join up with one another. How much will, say, telephone companies charge to carry entertainment channels (or will the balance be the other way around)? Will cable-television companies charge mobile-telephone operators or will both sides, as with Internet connections, work on a "sender-keeps-all" basis? Deciding who pays what, and whether it is fair, will be even more complicated when different services need different amounts of bandwidth. When Bell Atlantic or BT delivers the Disney channel across its local telephone loop, one thing is certain: it will not charge Disney the same way it charges AT&T for delivering telephone calls.

Rules on numbering

Interconnection breeds many rows. But some of the fiercest fights center on telephone numbers. Because numbers are a scarce resource (usually, no two customers have the same number), numbering systems provide all sorts of opportunities for anti-competitive behavior.

Dialing codes present one issue. In countries with competition in the long-distance market, customers need to dial a particular code to connect with the long-distance carrier of their choice. Dialing a long-distance number using, for example, Japan's giant NTT requires no special access code, but dialing through one of the three smaller competitors (DDI, Japan Telecom, and Teleway Japan) requires use of four extra digits.

Number portability constitutes another issue. In most countries, the big telephone companies decide who gets what number. When new telephone companies set up business, they soon find that changing telephone numbers deters potential customers. Hong Kong's telecoms watchdog calculated in 1995 that even tariff discounts of 10 to 15 percent were insufficient to compensate customers for the nuisance and expense of changing their numbers.[16] So who should pay the cost of allowing customers to keep their numbers, the newcomers or the old giant? Many countries have barely begun to think about these issues.

Both interconnection and numbering conflicts arise from the terms under which big networks interconnect with small ones. Such rows will die down once several networks of roughly similar market strength coexist, all in position to gain from interconnection, as where, for example, mobile telephones interconnect with fixed-wire services.

Is it possible to have a stable system with several rival networks, or will the rivals tend to amalgamate into a single giant? The evidence from other networked industries mildly encourages belief in stability: for example, in the past, countries had several railroad networks and canal companies; today, the United States has many competing airlines. But, given the tendency of networks to concentration, governments that want to ensure thriving competition may need to step in and give it a helping hand.

Giving new entrants a hand

In most industries, antitrust regulators try to ensure that all companies can compete on more or less equal terms. In some parts of the communications industry, that is difficult. In the telephone business, for example, would-be competitors face two challenges: an existing giant and economic pressures for concentration.

Is it enough to apply rules equally to both new entrants and the erstwhile monopoly? Or should the terms be deliberately asymmetric, with heavier obligations laid on the incumbent giant than on the newcomers?

Put like that, the answer may sound obvious. But asymmetry brings at least two further difficulties. First, the new entrants may not be minnows. In the United States, the regional Bell operating companies grumble about having to make concessions to help AT&T into their home market; in Germany, Deutsche Telekom complains that its new rivals are backed by some of Europe's biggest utilities and retailers; and in Britain, the cable companies that provide telephone services in competition with BT are backed by American companies such as Nynex and TCI.

Second, when competition succeeds, it drives down prices. But the incumbent giant will more likely be able to withstand the financial consequences of falling prices than are new entrants, which may have been betting on the shelter of big margins to earn their own almost-as-big margins. In Japan, the government's decision to allow NTT to compete in its market has terrified KDD, the main international telephone company; in Britain, the government has found itself trying to shelter the returns of cable-television companies. "They are building an alternative local telephone network," goes the implicit bargain, "so we should limit the risk to their investment." It is extremely awkward when new entrants owe their survival to government-imposed constraints on their established rivals.

Third, when should asymmetry end? A regulator who starts giving special favors to one group of companies may find it difficult to decide when and where to limit them.

An alternative to asymmetry would be to divide the provision of infrastructure from the provision of services, forcing companies to offer one or the other. One company (or, perhaps, several) would construct, operate, and maintain the network, or at least the local part of the network; others would be free to offer services across it. A cable company, for example, would have to choose between making programs or owning cable systems.[17] Sweden tried a scheme of this sort, with Stockholm's city council setting up a company to run fiber-optic cable through its many ducts, with the aim of leasing out capacity to anybody willing to pay for it. The test of such a solution will be whether such a divide, separating infrastructure from services, discourages innovation—or whether it puts different content producers on equal terms.

New Media, New Monopolies
· · · · ·

While the telephone and cable-television companies fight rearguard actions to protect their franchises, other industries in the communications business have created companies that control 70 percent or more of their markets: companies such as Microsoft, Intel, Cisco, and (for a time) Netscape. Many people worry that these will be the AT&Ts or local cable monopolies of the future. Microsoft in particular has been much investigated by trust busters—not surprisingly, since Microsoft represents by far the most important and enduring example of private control of a technical standard.

Microsoft's assault on the market for Internet "browsers" increased such worries. Netscape, after giving its browsers away free, began to charge fifty dollars apiece for the program. Microsoft stepped in with an equivalent product and a pledge to keep it free and to build it into Windows. Netscape's market share dropped steeply.

Some economists see a further cause for concern over communications and media monopolies. They believe that the concentrating tendencies of networks in industries where the costs of manufacturing and distributing extra products are infinitesimally low or even zero will lead to corresponding—and deleterious—concentrations of money and power. Brian Arthur, the leading exponent of this view, argues that the result of "rising returns to scale" will be a natural tendency for the market leader to get further ahead, causing a monopolistic concentration of business.[18] As a result, giant monopolies will increasingly dominate the new electronic world.

But the best is only the best until something better comes along. In high-technology industries, the astonishing pace of innovation means that market leaders must keep innovating if they want to stay ahead. Microsoft, the company that most people associate with rising returns to scale, has been an energetic innovator, ready to change tack quickly if it has mistaken the direction in which a technology will drive the market. The Internet boom caught Microsoft by surprise; but the company responded quickly, moving into Internet products with its own browser and family of on-line services.

Today's threatening monopoly may be tomorrow's sickly giant. Who cares now about IBM's market power? As for Microsoft, it faces a potential challenge from Java, a programming language that may make users less dependent on Microsoft's operating system and from the network computer, which could challenge the PC. In the fast-changing world of cyberspace, launching good new products faster and cheaper than ever before is the best defense against another company's dominance.

Regulators rightly worry about abuse of dominance, not about dominance itself. Two regulatory approaches make abuse less likely. First, faced with a dominant player, old or new, regulators should lean toward making entry into the market easier for new players. At present,

Microsoft's operating system is "open": it makes public the codes that allow competitors' applications to run smoothly on Windows. In telephone industry terms, this is the equivalent of allowing easy interconnection. Such openness should be an automatic obligation on every dominant owner of a network business, from BSkyB to Bell Atlantic.

Second, regulators should also keep open the possibility of splitting delivery and content, or, in Microsoft's case, of forcing a separation between making operating systems and making applications. Such divisions may discourage innovation to an extent, but they are sometimes essential if large players are to be prevented from abusing their market power.

As a general rule, regulators do better to err on the side of doing too little, rather than to risk doing too much. Even well-meaning regulators may fail to see which interventions will be good for competition and which will not.

The right approach in every part of the communications industry should be the least regulation that is consistent with ensuring lively competition. The complexities of interconnection, in particular, will always require a specialist arbitrator to settle disputes. Much else should, in time, be a matter for ordinary antitrust regulation.

For many countries, the fastest way to foster competition will be to strip away all controls on which industries can offer which communications services and on foreign ownership. The more competition emanates from unexpected directions, crossing conventional industry divides and leaping national borders, the more innovation and price competition consumers will be likely to see.

Policing the Electronic World

Orthodox Jews are among the many religious people who recognize the importance of the Internet but regard it with suspicion. Toranet, an Israeli company, has responded by building an Internet service that gives subscribers access only to specially screened Web sites. The Orthodox can thus be sure that anything remotely offensive has been filtered out, including all pictures of women, even those modestly dressed. On the other hand, they can find sites with information on sex-segregated swimming pools, kosher restaurants, and the times of the beginning and end of the weekly Sabbath.[1]

As distance shrinks and communications channels multiply, the problem to which Toranet provides an answer will only grow. As entertainment and information easily hop across borders, one culture's standards and morality will intrude into the homes of people living in other cultures with other standards and morality. Satellites beam television programs across national frontiers; telephone services such as sex chat lines can, with the aid of a computer and some fancy switching, move from country to country without a caller noticing the difference; the Internet allows people to look at on-screen information coming from anywhere in the world.

Electronic media embody information and ideas. They thus have extraordinary power to liberate, corrupt, subvert, or enfranchise, depending on how you look at it. Ithiel de Sola Pool called them "technologies of freedom," arguing that they create an unprecedented opportunity for freedom of speech, which should be reinforced rather than constrained.[2]

But all societies constrain free speech in some ways. Even in the most democratic societies, laws exist against the nastier sorts of pornography, libel, fraud, and terrorism and to protect brands and ideas. These abuses of free speech harm others. Such laws and their enforcement are generally local or national matters, however, and the new world of communications is global.

Three kinds of policing issues dominate debate about protecting people in the new electronic world: free speech, privacy, and intellectual property. The electronic world makes communication easier than ever: how far is it desirable—or even feasible—to regulate what people say electronically? Electronic media make it easy to collect data on people: how should that power be controlled? Electronic media distribute ideas and information: how can the value of these assets be secured?

Dealing with these issues is complicated in two ways: first, by the border-hopping character of new communications; and second, by the fading of the line between public and private.

Increasingly, countries will find unwanted material crossing their borders—material that is seditious or offensive to local taste or culturally threatening. National laws against such materials will be difficult to enforce against electronic transmissions. Tracing the sources of such material, for example, can become more difficult than in the case of paper, making jurisdiction unclear. Services can migrate to countries where laws are lenient or weakly enforced, creating offshore havens for pornography, gambling, tax evasion, and breaches of international rules concerning intellectual property.

Such border-hopping helps to explain why many countries regard the Internet in particular as a threat—to their culture, their moral norms, their religious scruples, and their political sensitivities. France's President Chirac, alarmed by yet another transatlantic blow to the French language, may have been resorting to Gallic hyperbole when he described the Internet as "a major threat to humanity,"[3] but people in many parts of the world feel qualms about the difficulty of controlling what comes down the wire or over the air.

In the United States and a few other democracies, the debate has been more about whether government should control the way communications media are used rather than whether governments can intervene to control content. The United States is unusual in giving a constitutional guarantee to freedom of speech. In addition, Americans mistrust their government more than do citizens in most other democracies. Because the Internet remains as yet largely an American colony, American values tend to shape it: its content is thus less likely to jar on

Americans than on others. But, even in the United States, many people worry about the effects on children of easy access to pornography and violence, about intrusions into personal privacy, and about the ease with which criminals can have access to on-line manuals on bomb-making and terrorism.

Disagreeable content is not the only problem in the electronic world. Information in digital form can travel over several different kinds of networks, to be received on different sorts of screens. In the past, governments in democratic countries have rarely concerned themselves with what their citizens said to each other in private, in letters, or over the telephone. They have, however, taken a great deal of interest in information that appeared in books or newspapers, on the radio, or on television. There have thus always been two classes of intercourse: one private and largely unregulated and the other public and subject to rules about matters such as decency, copyright, libel, and, in some countries, political balance. This divide allowed governments to impose one set of rules on broadcasters, say, and another, less restrictive set on individuals. Things can be said freely in private that could never be repeated on a television program.

Now, the line between public and private is blurring. The same network can deliver a newspaper, a broadcast, or a private letter; the same terminal can receive all of these. What starts out as private—an e-mail, say—can move seamlessly into the public arena (by being posted on a Web site, for instance). As a result, arguments about all sorts of aspects of electronic communications—copyright, sexual harassment, inde-cency, and so on—quickly turn into debates about free speech. Attempts to regulate the former have inescapable effects on the latter. Public and private space, electronically speaking, will be harder to separate.

The blurring of public and private affects not just the policing of con-tent but also the protection of privacy. It is quite true that "On the Inter-net, nobody knows you're a dog," but many people do know a great deal about your on-line interests, canine or otherwise. Governments and companies now know more about the pattern of an individual's life than ever before. People electronically connected to an interactive network—be it a telephone network, cable or satellite television, or the Internet—can be monitored in all sorts of ways. Any company that manages subscriptions and bills customers has a huge amount of detail about their tastes, incomes, and spending patterns. Such information, when stored on electronic databases, will be potentially accessible to governments, employers, and hackers. An Orwellian regime, determined to control its citizens' private lives, would have technological tools to monitor their activities beyond the most sinister dreams of Big Brother back in 1984.

A world without distance will require a different approach to the diffusion of undesirable and private information. Government control will be possible, but only with a greater intrusion into personal freedom than voters in democracies are likely to tolerate. Instead, governments will have to find other ways to discourage anti-social behavior. And sometimes they will simply have to accept that, if people want to behave badly in ways that do not harm others, society will find it less costly to allow them to do so than to try to prevent them.

When Regulation Is Desirable
. . . . .

Since it grew up among American academics, the Internet, not surprisingly, acquired a culture that values free speech. After all, one of its great boons has been to make accessible information once available only to the elite. Many Internet regulars, especially Americans, would argue that the freedom to argue and discover is worth defending at all costs, even if its corollary is the freedom to titillate or abuse.

In the rich countries, most of the argument about censoring electronic communications has been concerned with preventing children from watching pornography and violence or with stopping some people—swindlers, racists, terrorists—from communicating information that most citizens regard as undesirable. In some other countries, the primary issue involves trying to prevent adults from reaching material—such as political criticism—that they want to see.

The availability of child pornography on the Internet is particularly disturbing to many people around the world. In the United States, in September 1995, a two-year investigation by the Federal Bureau of Investigation ended with the arrests of dozens of people for using America Online to distribute pornographic pictures of children and to solicit sex with minors. In Britain, police believe that the Internet has made child pornography available to people who a few years ago would have had no idea where to find it.[4]

Not only can anybody with a personal computer and a modem become a global publisher of pornography, but anyone can just as easily make available racist and terrorist materials. Instructions for building an ammonium nitrate bomb of the kind that killed 167 people in Oklahoma in 1995 were found on the Internet. In spring 1996, France asked the United States to crack down on an Islamic group in San Diego that had been putting on the Internet instructions for assembling inexpensive bombs such as those that had recently exploded on the Paris metro; American officials reportedly offered "sympathy" but little else.[5]

If such things upset people even in the rich and relaxed west, how much more is there to upset people from countries with sterner cultural standards? What passes without notice in one country may be considered racism or blasphemy elsewhere. Catering to individual national sensibilities is relatively easy with television, which is mainly distributed separately to individual national markets. With the Internet, it is not.

In Sweden, for instance, owning pedophilic material is not illegal (although it is an offense to transmit or market it); in parts of the Middle East, a picture of a girl in a bikini is an outrage. Such differences in national outlook extend to politics, as well. One country's terrorist is another's freedom fighter. So people in the west may be delighted that Chinese dissidents used the Internet during the Tiananmen Square demonstrations; or that Bahrainis use it to rail against their repressive royal family; or that Serbs used it to stir up opposition to the government of Slobodan Milosevic. They may feel less happy when America's white supremacists use it to plot racist crimes.

The history of censorship demonstrates the immense difficulty of defining by law what should or should not be proscribed. Are pictures of naked women an art form when painted by dead Italians but pornography when photographed by live ones? Even cohesive societies with uniform religious or cultural values find it difficult to decide where to draw lines; for a community of perhaps seventy million people in 160 countries, the task is infinitely harder. But while that argument works against the idea of trying to set global standards for electronic content (a task that would be doomed to failure anyway), it does not exempt Internet users from adhering to whatever set of rules countries choose to apply to maintain decency and oppose libel, child abuse, gambling, terrorism, tax avoidance, racial abuse, political bias, advertising, or anything else. Simply because the Internet is a new technology does not mean that it deserves a new set of rules. Existing rules may need to be adapted in order to apply, but their basic goals need not be altered. Even the most ardent nethead would presumably accept that a fraud is still a fraud even if perpetrated on-line: why should a breach of public decency laws be any different?

But the philosophical principle—that the same rules should apply in the electronic world as apply in the real world—constitutes only a starting point for answering the policing question. Beyond lies a tougher question: can any rules be made to work?

Whether Regulation Is Feasible
• • • • •

With television and the telephone, censorship can be applied relatively straightforwardly, if a government is determined to have it. In the case of television broadcast across borders, governments can always outlaw advertising or the collection of subscriptions if they dislike the programs. Rupert Murdoch, having boasted about the ability of satellite television to topple tyrants, has found it necessary to adapt the content carried by his Star Television service in China. To appease the Chinese government, Mr. Murdoch took the BBC World Service's television broadcasts off Star's satellite.

Telephone calls are much more difficult to monitor, mainly because it would take an army of listeners to do so. But, in the case of sex-chat lines, regulators have a point from which to apply pressure: the service providers need to advertise and collect payments. When, for instance, an operator of sex chat-lines transmitted from Guyana to Britain broke the rules by advertising its services in a family newspaper, Britain's regulator threatened to route all calls to Guyana through an operator. Guyana promptly closed the line.[6]

With the Internet, though, censorship becomes more difficult. For censorship to work, it has to resolve three issues. First, what line should the law draw between the public and the private? Second, does the technology exist to screen out offensive material? And third, who should have the legal responsibility for doing the screening?

The line between public and private

The United States opened a debate on the question of censoring the Internet with the February 1996 Communications Decency Act, passed by Congress in a rush as part of the telecommunications bill. The law intends to make it a criminal offense for a person knowingly to transmit "obscene or indecent" material to anybody under eighteen; to display such material "in a manner available to" a minor; or to allow use of "any telecommunications facility under his control" for such prohibited activities. Such sweeping stuff brought a number of groups, including the American Civil Liberties Union, rushing to test the law in court as a possible violation of the guarantee of free speech in the first amendment of the U.S. Constitution.

As a result, a 1996 landmark court case in Philadelphia decided, to the delight of the Internet's devotees, that the Communications Decency Act was unconstitutional, a ruling confirmed in the Supreme Court in 1997. The Philadelphia judges argued that the Internet represented "a never-ending worldwide conversation."[7] Where the law decides that the Internet is essentially a private medium, it will be bound to regulate it more leniently than a public medium.

Suppose, though, that the law treats the Internet as a public medium. The key difficulty then will be to curb the production of offensive material. Liability for publishing legally unacceptable material would normally rest with the publisher: the person who put the material on the Internet in the first place. But on the Internet, that is clearly unworkable. Material may have been made available by somebody in a different country; it may be stored on computers scattered around the globe, moving across many national borders on its way to the end user.

Besides, because it is so inexpensive to put material on the Internet, much of the sinister stuff is the work of individuals who want no commercial reward. With television, as Mr. Murdoch's experience in China shows, governments have an obvious target: even a cross-border broadcaster usually hopes to collect rewards from national subscriptions or national advertisers. With the amateur on the Internet, no such leverage exists. The question becomes whether undesirable material can be filtered out before it reaches the user.

. . . . .

The technology of screening

The answer as to whether screening can be effective matters both to dictators and to parents. The same screening techniques, broadly speaking, that allow China to restrict access to the technologies of freedom allow parents to keep their little ones from looking at nasty pictures on the Web.

But there is a difference. Screening carries a cost. For families, censoring themselves, that will not matter. Where governments censor unwilling citizens, however, not only will it be harder to make screening work, but the costs, in terms of useful information forgone, will be greater.

China or Saudi Arabia or Myanmar care little for such costs. They limit what most of their citizens can watch on satellite television or read on the Internet because they are prepared to take measures far more brutal than any open society would contemplate. Thus China insists that all the country's Internet users channel electronic communications through a series of filters monitored by the Ministry of Posts and Telecommunications and other agencies. All domestic Internet users are obliged to register with the police, and companies and individuals using or providing access to the Internet are "forbidden to produce, retrieve, duplicate, or spread information that may hinder public order."[8] China aims to create a sort of giant national intranet (which Sun Microsystems is helping to build on a $15 million contract): a self-contained network linked to the main global Internet by means of a narrow channel that can be closely monitored and controlled.

The practical problems—and the costs—of censorship increase in a country such as Singapore, which wants to build its economy on the quality of its electronic communications. In September 1996, Singapore required all individual subscribers to route requests for access to Internet sites through "proxy servers," large computers that check such requests against a blacklist of sites containing objectionable material—mainly pornography and political criticism of the regime. Requests for banned sites are blocked.[9]

The government does not require businesses to go through the proxy servers, and indeed it would be hard to run the system if they had to do so, because the software shuts off access to entire Web sites rather than to individual pages. Such an approach would be far too costly for a sophisticated economy to justify inflicting on itself. Singapore's censorship, therefore, remains only partly effective, a casualty of the dilemma of a government that wants to be both safe and modern.

Interestingly, other advanced Asian countries have shown little desire to emulate Singapore. They do not necessarily possess greater enthusiasm for free speech: Malaysia proscribes Singapore's newspapers, even though any Internet user can read them as soon as they are published. But Malaysia's grandiose plans for a "multimedia super-corridor" near Kuala Lumpur, with the goal of being what Mahathir Mohamad, the prime minister, calls "the world's best environment for multimedia business," sit uneasily with censorship. The Malaysian government has therefore decided that industries located there will enjoy a more relaxed approach to content regulation than does the rest of the country. Developing countries that want to attract successful multimedia businesses will find that they cannot afford the luxury of censorship.

Where countries are determined to screen content on the Internet, they must choose one of four main approaches.[10] The first approach uses software designed to consult lists of known proscribed sites and to block them according to criteria chosen by the censor, by an organization chosen by the PC owner, or by the PC owner. This approach depends for success on having a complete and accurate list of undesirable sites, which becomes a problem, given the huge number of sites. An assiduous censor could easily block the obvious Web sites, such as Playboy; but keeping track of the ever-changing hordes of less obviously objectionable ones is more or less impossible. Any group that wants to make the censor's life difficult can always call itself something misleading.

The second approach uses software that looks for taboo words as a basis for deciding whether to block a given site. Computers, however, are bad at distinguishing meaning. America Online caused a furor when its word-screening software shut down a forum for discussing breast-cancer because it mentioned breasts; software from another company closed off access to the official White House site because of a reference to the presidential "couple."[11] Foreign languages and pictures (no computer has yet learned to tell a dirty picture from an innocuous one) create yet more difficulties with this censorship strategy.

The third approach requires Internet service providers to screen off objectionable areas of the Internet. Singapore, China, and, indeed, Toranet use this option. Blocking access to newsgroups is relatively straightforward: unlike Web sites, newsgroups broadcast to each service provider's computers. The service provider can at that point easily screen them—or shut them out. Looking at the content of every newsgroup would be impossibly labor intensive, but censors can always instruct service providers to allow access only to approved newsgroups.

Dealing with Web material presents a tougher problem. Web sites remain on distributors' computers for viewers to "visit." If governments want to keep viewers away from particular sites, service providers can be asked to close them off. But the site or bulletin board may reappear, hydra-like, if necessary under a different name. When German prosecutors tried to shut down a California-based site run by Ernst Zündel, a German neo-Nazi living in Canada, their efforts were thwarted by defenders of free speech who posted copies of the material elsewhere. Because such material is illegal in Germany, Deutsche Telekom, the largest provider of Internet access in Germany, shut off all access to the computers of the company that had rented space to Mr. Zündel. This blunderbuss approach deprived Germans of access to another fifteen hundred unobjectionable sites as well.

A fourth approach constitutes a form of self-censorship. In this approach, governments persuade content owners, or third parties, to rate sites voluntarily, just as films are rated and many computer games and videocassettes are tagged. Electronic tags placed on sites would warn families of unsuitable material (so that the PC owner, or, for those who had requested the service, the Internet service provider could block it automatically). Tags could also be used to opt into services: ISPs could promise families only approved material. In a competitive market, parents might choose to subscribe to an ISP that offered censorship.

In the spring of 1996, specifications for the Platform for Internet Content Selection were agreed upon by a group of people from Internet companies and organizations. This agreement provides for one of a number of services (run by a parents' group, for example, or a specialist company) to rate Internet sites according to specifications set by parents (or employers or a service provider) describing prohibited sites; the browsers on subscribers' computers will not download a forbidden file or allow a visit to any unrated site.[12] This effectively deals with the problem that what outrages one person may not disturb another. Such a system puts censorship where it is most likely to be effective—in the hands of the individual citizen.

In time, some such system will probably extend across television too, allowing a common ratings system with a choice of categories and application for both media. Already, software is being used to screen out television programs that do not meet a certain rating. A law passed in the United States early in 1996 requires all new television sets to be fitted with a "V" chip—a circuit that will allow parents automatically to block programs that they do not want their children to see.

All such systems, however, share a basic problem (quite apart from the fact that most twelve-year-olds are more adept at manipulating technical gizmos than is the average middle-aged parent): the ratings system will be no better than the labeling service that decides whether particular content is offensive. The television industry rates programs according to their supposed appropriateness for children of different ages; it does not rate news programs, which may contain scenes even more disturbing to children than fiction because they know it is real. The Internet will most likely give rise to competing ratings schemes, allowing families a choice.

But voluntary ratings will not cover material that comes from abroad. Screening all available sites on the Internet is an immense task. Ultimately, the only sure-fire way to control what people see will be by opting in, rather than leaving out: by restricting what can be viewed to a limited number of safe sites or approved television channels. For rich-world families trying to protect their young, that will mean voluntarily cutting off access to much of the Internet, a sacrifice they may feel it is worthwhile making.

Who should be accountable?

Self-censorship is the ideal way to control electronic unpleasantness. If parents care enough about what their children see, on television or the Internet, the market will devise ways to ensure that families receive sanitized material only. But this is not a complete solution. It may protect children, but it does nothing about material that would be illegal if distributed in other ways.

In the search for leverage, regulators everywhere have tended to target the Internet Service Provider or ISP—the company providing Internet access—as the one point at which they can apply pressure. One of the most notorious attempts to do so was made in Germany. In December 1995, a prosecutor in Munich, Bavaria, told CompuServe that about two hundred Internet discussion groups on sex-related topics violated German law, and particularly laws designed to protect minors. Because CompuServe had no technical means of tailoring its Internet content (which is held mainly on computers in Ohio) for German subscribers, it had to apply the exclusion everywhere in the world. Access to the offending newsgroups was denied to all of CompuServe's four million subscribers worldwide. In April 1997, the Bavarian state authorities went further and charged the managing director of the German division of CompuServe with providing access to pornographic and racist material on the Internet.

The CompuServe case attracted much attention because it appeared to imply that the standards applied by one country (or state—the ruling was made under Bavarian law) could be enforced right across the Internet. In fact, CompuServe could easily have chosen simply to pull out of Germany (or to ignore the ruling). If Germany had been Guinea, and the complaint had been about carrying material on some disagreeable African potentate, presumably the company would have shut down its national service. But CompuServe clearly deemed the German market worth the fuss: cutting off Germany to protect a few smutty newsgroups would hardly have enhanced the corporate image.

Several governments, and several American states, have tried to make ISPs responsible for anything obscene that goes through their lines. The ultimate legal question becomes whether to treat ISPs like broadcasting companies or like telephone companies: are the ISPs common carriers, acting merely as conduits for messages, or are they publishers?

The U.S. Communications Decency Act, passed in 1996, implied that ISPs are not sheltered by the common carrier defense from responsibility for the content they provide. But without some such defense, ISPs face a Herculean task. The decision in the Philadelphia court challenge of the act took a different view. To a much greater degree than for a television broadcast, the judges pointed out, material on the Internet has to be deliberately sought out by the viewer. A child sitting in front of a television screen may find herself unexpectedly watching an inappropriate program, but a child sitting in front of a computer screen is at no such risk. As one judge put it, "The Government may well be right that sexually explicit content is just a few clicks away from the user, but there is an immense legal significance to those few clicks."[13]

German courts may eventually come around to a similar view. In April 1997, while Bavaria was pursuing CompuServe, Germany's parliament was discussing a new law to put responsibility for illegal content with those who create it. "Providers that only transport contents can't be made liable for these foreign contents," said Edzard Schmidt-Jortzig, the Federal Justice Minister. "That is only the logical consequence, because we don't punish the postal service for transporting letters with instructions for concocting Molotov cocktails, Nazi propaganda, or child pornography. Those punishable are those who send it—whether over the Internet or as a letter."[14]

This debate presents no easy solution. Even China, with virtually unlimited supplies of labor, will probably not succeed in keeping the most determined users of the Internet away from undesirable content. Anybody anywhere can dial up a service provider in another country, the main drawback being the cost of an international rather than a local call. Stopping such access would require cutting off international telephone service. Even China might hesitate to take such a step. Certainly it is out of the question for an industrial democracy.

Rich countries will simply have to decide how far to push the point. It will be impossible—and unfair—to force ISPs to take blanket responsibility for what their customers can reach on the Internet. But many governments may decide to oblige ISPs to take responsibility for material they knowingly make available to their customers. The key test will be to define what constitutes "knowingly," and such regulation will entail a trade-off: the more easily and frequently ISPs are prosecuted, the more costly it will be in terms of both money and access forgone. And, despite every effort, the determined user will still be able to seek out and find socially objectionable materials.

Protecting Privacy

. . . . .

Paradoxically, the electronic media make it easier for pornographers, hackers, and swindlers to hide behind anonymity while at the same

· · · · ·

time representing a serious threat to privacy. Indeed, the Internet links the two issues: the more regulators intervene, or force ISPs to intervene, to control who looks at what, the more will be known about users' habits. The evil Mr. Hyde can easily claim on-line to be the charming Dr. Jekyll—but whenever he uses interactive media he will leave electronic footprints behind. Nobody may know his name, but all sorts of people will know his tastes.

Electronic media thus offer more ways to accumulate data, raising two fundamental questions about privacy. Is the accumulation of private information a potential threat to liberty? And what controls should protect the use of data?

The accumulation of data

More and more facts about people's habits are now squirreled away on databases. Many have been collected by governments, in an attempt to govern more efficiently: databases help to tackle traffic jams, track fraudulent welfare claims, monitor book loans from libraries, and, of course, catch criminals. The state of Maryland, for example, requires that hospital visits be logged on a database in order to make health care administration more effective; and in 1996, five states ran a pilot program under which employers hiring a worker had to contact the federal government to determine whether the person was an illegal immigrant.[15]

But the culprit is not so much Big Brother as lots of little brothers. Make a purchase with cash, and you may vanish into the crowd; make it with a credit card or check, and you will increasingly end up on a mailing list or ten, deluged with brochures offering products and services distantly related to the object you have just acquired. Americans are on more databases than any other people on earth, partly because they make more payments with a credit card than do people in other countries, and therefore leave a more detailed data trail, and partly because the United States stores far more information on far more computers than do other less computerized nations.

All three communications technologies contribute to the growing data trail. Two sorts of intrusion have become easier with the development of the digital telephone network: call records and caller identification. Call records are routinely used by law enforcement agencies as evidence in trials to show who called whom, where, and when. Such records are also used in civil actions, such as the wrongful death suit against O. J. Simpson. In Britain, records of calls to Malaysia and Indonesia, where match-rigging is rife, were an important part of the prosecution evidence in the trial early in 1997 of three footballers charged with fixing matches. Also used in the trial was data on where their mobile telephones (switched on, but not in use) were taken and for how long.[16] And in April 1996, Dzhokhar Dudayev, leader of the Chechen separatists, was apparently killed on a hillside while making a telephone call. He was hit by a missile fired from a jet after the Russians pinpointed his whereabouts by intercepting the signal of his satellite phone.[17]

Using caller identification, not only can individuals tell who has rung them, even if the call is not answered, but so can a host of companies. Ring a toll-free or premium-rate number and the company can note your number and, if it wishes, sell it to another firm.

Pay-per-view television presents the opportunity to collect precise information on what programs people watch. Oddly, in the United States one of the few bits of information about customers that is protected by law is what movies people rent from a video store.[18] Perhaps not surprisingly, many of the trials of interactive-television services have emphasized transactions systems, used to sell and to monitor what is sold, rather than entertainment.[19]

The Internet undoubtedly adds to the amount of information that can be collected about those who use it, as anybody who has looked themselves up on screen can testify. It offers three legitimate ways to collect detailed information about users. First, many sites offer free content to users—in exchange for a large amount of detailed personal information. Such information is, of course, only as useful to advertisers as the provider is honest. One British user found that, by giving his birth date as that of the Duke of Wellington (1 May 1769), he was deluged with offers of special senior-citizen discounts from airlines.[20] Second, servers can increasingly store information about who visits a particular site. Almost every computer connected to the Internet has a file (at "cookies.txt") that keeps an automatic record of which sites the user has visited. Eventually, this file will also record how long the user spent there and any purchases made. This can be to the visitor's convenience: users who, say, regularly start by looking at the Asia section of the on-line *Economist* will soon find that the site offers them Asia straight away. This function can also allow more precisely tailored advertising. But it clearly has big implications for privacy. Third, customers who buy products over the Internet will find their purchasing decisions, like many credit-card purchases, recorded for future use by the junk-mail industry.

Privacy issues also affect relations between companies and their employees. Software exists to allow employers to screen employees' electronic mail (for example, to pick up a word such as "resume" or "cv"), and such screening is legal in most countries. Software also exists to allow employers to monitor the sites employees are visiting.

The potentially invasive use of data gathered through Internet use gives new urgency to the old debate about the balance between protecting the privacy of the employee and protecting employers from repercussions of illegal or undesirable activities by their staffs. The Internet increases the scope for both parties for both intrusion and abuse.

A threat to liberty?

By collecting data, both governments and companies operate more efficiently. Governments, for instance, can check how efficiently they provide health care or education. Companies fretting that they waste half their marketing budgets but do not know which half will (theoretically) be able to aim their sales pitches more precisely. A well-organized bank will no longer try to sell you life insurance each time you telephone if it succeeded the first time you called.

But many people worry about the collection of so much personal information. Are they right to do so? Peter Huber, an American communications expert, has argued forcefully against the idea that electronic communications allow government to threaten a citizen's privacy. In *Orwell's Revenge*,[21] a powerful rebuttal to George Orwell's vision of the future in *1984*, Huber argues that the sheer labor of keeping check on millions of citizens would, of itself, deter Big Brother from doing so. Maintaining a communications network, he says, requires the cooperation of too many skilled people for any totalitarian government to monopolize the task; indeed, the interactivity that

Orwell envisaged would allow citizens to talk to each other and so to arrange the overthrow of Big Brother.

Mr. Huber's views seem persuasive when applied to established democracies with long traditions of respect for civil rights and the rule of law. Indeed, it is no accident that the countries where copious electronic databases have been compiled have been those where governments are most scrupulous about protecting civil liberties. Americans part with large amounts of information about themselves because they are reasonably confident that it will not be misused.

But Mr. Huber's view ignores two points: a government may want to keep track, not of millions of citizens, but of a few; and computers will readily spot the few in the crowd. In countries where civil rights are weak, interconnected electronic databases will make some aspects of government control easier. In December 1993, to take one sinister example, China announced a plan to carry out genetic screening of its population by the end of the century. Women carrying defective genes would be sterilized or prohibited from marrying.[22] Such intrusion may seem unimaginable in the rich world—until you substitute insurance companies for the government or wonder whether Medicare should go to lifelong tobacco-smokers.

Western democracies face the real danger that personal information compiled for one purpose will be used for others. By cross-referencing information stored on databases, governments or commercial organizations may learn too much about an individual citizen. In addition, any database connected to a network becomes ultimately vulnerable to hackers. Most disturbing of all, search services have been created that allow ordinary people to discover large amounts of information about others. Electronic mail addresses are readily available, as are ways to search through the Internet for references to individuals. P-Trak, launched in 1996 by Lexis-Nexis, one of the biggest providers of on-line data, originally allowed access to an individual's name (and maiden name), telephone number, current and two previous addresses, and month and year of birth.[23]

Such intrusions may be qualitatively no different in kind from steaming open envelopes or listening in on telephone switchboards. But computer networks allow the well-organized snoop to ferret out information more efficiently than ever before. At the least, the collection of data calls for tough controls on its use and resale.

Protecting personal data

National attitudes toward privacy differ enormously, reflecting the extent to which citizens in different countries trust their governments and private industries. Americans assume that the Supreme Court protects their right to privacy: in a famous 1965 case, the Supreme Court found the right to privacy to be protected in the "penumbras" and "emanations" of the Constitution. This has led many Americans to believe that data collected for one purpose cannot be used for a different purpose without the consent of the person concerned. In fact, that right has effectively gone by the board. Instead, legal restraints on the sale of personal information are less stringent in the United States than in some other parts of the world.

This is an area where national attitudes vary widely. In Sweden, a country with a relaxed approach to privacy, the government actively sells to direct-marketers information that it has compulsorily collected from its citizens. Companies in Sweden can, for instance, buy a Swedish citizen's unique identifying personal number and use it to acquire that person's complete car-owning and driving record or buy information on a citizen's address, family status (including the numbers of family members), income-tax statement, credit status, and the value and location of any property holdings. In addition, it is not the Swedish government nor even a Swedish company that manages most of this information, but the SEMA group, part-owned by France Telecom.[24]

At the least, the protection of privacy surely calls for firm controls on the use of information and on the extent to which information collected by one organization can be put at the disposal of another. Some such controls already exist in many countries. Americans, for instance, have a legal right to see most of the files kept on them and sometimes the right to correct mistakes and to block disclosure. But most individuals would find it difficult to track down such files and ensure that they were accurate. In the United States, people can register with the opt-out list of the Direct Marketing Association.[25] In Britain, people can sign up with the Telephone Preference Service, set up by the telecommunications and telemarketing industries to give consumers the choice whether or not to receive calls from companies with which they have never previously had dealings.[26] The European Commission may propose that every country in the European Union be legally required to set up such a scheme.

Individuals must be able to decide whether to allow collection of details on their lives, purchases, and activities. In the United States, where caller-identification has existed for longer than anywhere else, a long debate has explored the proper way to protect a caller's privacy. Callers can now choose to bar the transmission of their numbers to anybody except the operator and emergency services. But companies offering toll-free services automatically receive a caller's number, on the reasonable grounds that they are paying for the call. In addition, people should have some say in the reuse of information about them. Such information has value. Should that value belong to the individual or to the company that has collected it?[27]

These issues will become even more important as companies and governments sell information not just within countries, but across borders to countries with different laws on the use of information. In October 1995, the European Union agreed to a directive incorporating comprehensive data-protection rules. One provision of the directive bans transfers of sensitive information to countries with inadequate privacy laws. As personal details become an internationally traded commodity, people will begin to see them in quite a new light: as their intellectual property, perhaps, to be protected as carefully as a valuable brand or a unique idea.

Intellectual Property

· · · · ·

Electronic communications distribute ideas and information. These, increasingly, are the stock-in-trade of successful businesses. A good idea, whether caught in the form of a patent, a brand, or a movie, constitutes an immensely valuable commodity—an "intellectual property," an intangible with commercial value. The revolution in communications will make it cheaper and faster than ever before to distribute such property. But it will also make it cheaper and faster than ever before to reproduce intellectual properties without the permission of those who believe they own them.

The changes in global communications will stimulate the value of intellectual property in two main ways. First, global marketing tools, such as a single international toll-free telephone number or a Web

. . . . .

Table 7-1   Balance of Trade in Intellectual Property (Royalties and License Fees in Million Dollars for 1995)

Country	Total Fees
United States	20,660
United Kingdom	1,710
Sweden	–123
Mexico	–370
France	–470
Brazil	–497
The Netherlands	–700
Italy	–704
Australia	–775
Spain	–1,073
Germany	–2,660
Japan	–3,350

Source: Data from International Monetary Fund.
Note: Minus sign implies a net deficit.

site, give new value to global brands and trademarks. Merely national protection of these items of intellectual property will therefore be inadequate. Second, products such as technical papers, movies, and software—items of intellectual property—can now be distributed electronically. But with the ease of their distribution comes ease of unauthorized reproduction of copyrighted material, giving a new urgency to global, rather than purely national, efforts to prevent piracy. Both these points matter far more to the United States than to any other country. Not only does the United States have many more globally recognized brands than other countries, with names such as Coca-Cola, McDonald's, and Disney, but the United States is also far and away the world's biggest exporter of intellectual property, thanks to its enormous earnings from the movie, music, and software industries. (See Table 7-1.)

The battle for brands and trademarks

As companies publicize their products globally on the Internet, they run into a problem that sounds abstruse but is, in fact, fundamental to control of the intangibles of intellectual property: domain names.

Domain names, the Internet's equivalent of telephone numbers, are the distinguishing words that a company or other organization uses as its Internet address, such as www.economist.com. Given that, by May 1997, one million Web site addresses had been registered, a memorable, readily identifiable domain name, designed to stand out in the chaos and sprawl of cyberspace, is clearly essential. Anybody looking for IBM, for instance, would assume its address was www.ibm.com. A domain name thus assumes as much importance to a company as does a trademark. But, while a complex legal system protects trademarks (not always effectively) around the world, domain names have generally been distributed on the basis of first-come, first-served.

Not surprisingly, astute amateurs (and some smart small companies) have periodically pinched a name that a corporate giant regards as its intellectual property. A woman in Colorado acquired "rolex.com" before Rolex Watches of Switzerland realized its value; an American journalist grabbed "mcdonalds.com" before the eponymous hamburger empire. He had telephoned McDonald's to tell them that he was going to do so, but the company was not interested. After such well publicized cases, the American organization that issues domain names began to insist that claimants demonstrate a right to them. But that, too, is not a complete answer if two companies claim similar names. Roadrunner Computer Systems, a small Internet-service company in New Mexico, registered "roadrunner.com"—only to be challenged by Warner Bros., owners of a cartoon character called the Road Runner. Threatened by Network Solutions, Inc., the company that registers the main domain names in the United States, with having its site shut down, Roadrunner went to court. In mid-1997, its roadrunner.com site was still on-line.

While trademark law allows many companies legitimately to share a name in the high street or telephone directory, domain names can refer to only one user. The desirable permutations of useful names are finite and rapidly running out. Only one company can be registered as "sun.com," and Sun Microsystems happened to register it before Sun Oil or Sun Photo. As John Gilmore of the Electronic Frontier Foundation puts it, "Neither lawyers nor governments can make ten pounds of names fit into a one-pound bag."[28]

This problem, initially purely American, is becoming increasingly global, because of the desirability of having a name that ends in ".com." Most domains are grouped by country: a British company, for example, would end its name "co.uk," and a Japanese one with "co.jp." American companies rarely put a national tag at the end of their domain names. To be registered as "www.economist.com," therefore, suggests a global company, while "www.economist.co.uk" marks a business as a purely British concern. As a result, a rising proportion of names in the ".com" category do not designate U.S. companies. The official Chinese news agency, for example, has registered the name of "Taiwan.com" (to the indignation of the Taiwan government).[29]

In an attempt to solve these problems, a self-appointed international body of Internet veterans called the Internet Ad Hoc Committee proposed, in a report published early in 1997, the creation of more "top-level" domains with tags that are not country-specific such as ".firm" and ".store" as well as a system for adjudication of domain-name disputes. In addition, the IAHC suggested breaking the lucrative monopoly of Network Solutions by setting up a number of other registrars around the world.

This proposed solution came under attack from two directions. First, several governments and companies challenged the authority of the IAHC to tackle the problem. The trouble is that, with no central command, no single body has clear responsibility for deciding the Internet's future. Secondly and more specifically, the multiplication of tags will not necessarily solve anything. Companies with global brands, such as IBM and Sun Microsystems, will be likely simply to register under several domains.

If no solution emerges from the Internet community, governments may eventually take over responsibility for the domain-name system. In some countries, they are already being forced to intervene to deal with novel problems, such as the anti-trust implications of allowing a self-appointed group the right to charge for registering domain names. In Denmark, in spring 1997, some of the ISPs responsible for registering domain names changed the rules one night, grabbing for themselves the most valuable addresses which they then tried to sell to the companies and cities holding the names. Denmark's national competition authorities stamped on the scheme.[30]

Copyright and copying

Electronic media present new challenges to copyright holders. Copyright laws around the world were designed for an age when most copyrighted material was held in physical form: a book, say, or a CD. Copy-righted material converted into digital form can be copied perfectly without any damage to or diminution in the quality of the original. It becomes possible to create an infinite number of master copies. Electronic transmission adds a further problem: valuable products that can be transmitted on-line can also be illicitly copied. Indeed, the process of on-line distribution is simply copying. The Internet in particular can be considered one gigantic copying machine, distributing copyrighted material by reproducing it.

So electronic communications creates both special opportunities and special problems for companies and artists who make their livings from material that can be reproduced in digital form, such as books, magazines, music, movies, databases, and software. The main problems are the infinitesimal costs of reproduction, the difficulties of enforcing copyright, and payment systems. Behind these questions lies a more basic issue: can the concept of copyright survive in its existing form in an electronic world?

The costs of reproduction

The same economic force that turns copyrighted materials into potential money machines also makes them particularly tempting to pirates. Piracy constitutes a particular problem for the music and software industries. (See Figures 7-1 and 7-2.) Such products sell, not for what it costs to make a physical object (such as a book or a compact disc), but for a price that may reflect heavy research costs (as in the case of databases), an ingenious idea (such as movies or books), or spending on branding (as with rock groups).

They thus have high development costs and low production costs. Once a software program has been developed, millions of copies can be created and distributed on-line at no extra cost. This characteristic allowed id Software, producer of the game Doom, and Netscape, producer of the leading Web browser, initially to distribute their products free (see Chapter 4). But it also means that when such products are sold the price will be vastly greater than the costs of producing each individual product. Because it costs so little to produce a perfect fake, consumers and pirates share an interest in cheating the original creator of the product.

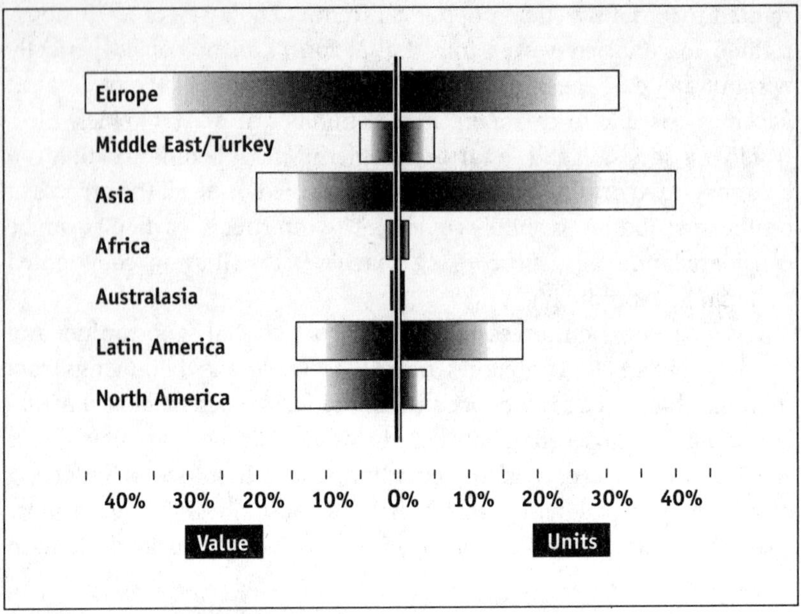

Figure 7-1   Music Piracy, 1995
Distribution of estimated music piracy losses and estimated piracy
rates by region.
Source: IFPI *Pirate Sales 95*, May 1996.

The difficulties of enforcement

At the same time that the temptation to piracy increases in the electronic world, efforts at prevention become harder, for four reasons. First, digital products are more prone to do-it-yourself piracy. When piracy means printing a book or manufacturing a compact disc, regulatory authorities have a hope of finding and raiding the pirating factories and distributors. Once copyrighted works can be distributed electronically and downloaded into an ordinary PC, any individual can, in theory, copy or alter any digitized work on the Internet and distribute it with a few keystrokes to hundreds of friends. As a result, violations of copyright become difficult to spot, and the law impossible to enforce. No government in a democratic country will routinely raid homes for pirated material in the way that police occasionally raid companies thought to be using or distributing pirated software.

Second, electronic media make it more difficult to establish a legal definition of copyright. One problem is that any visit to a Web site tech-

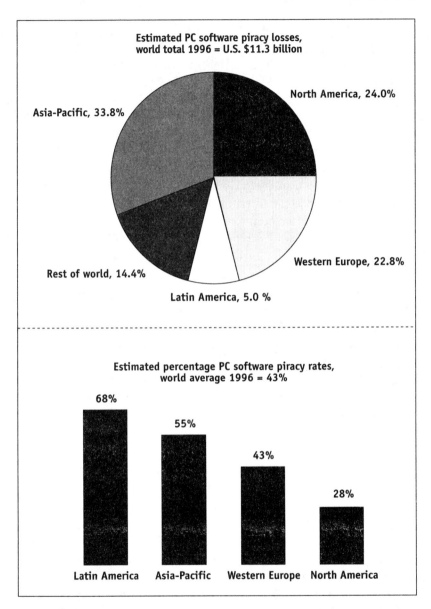

Figure 7-2   Software Piracy

Distribution of estimated PC software piracy losses and estimated piracy rates by region.

Note: Estimated losses of revenue on business applications only, at retail value. Piracy rate = number of applications pirated as a percentage of total number of applications in use.

Source: Software Publishers Association.

nically involves the automatic downloading of material, however temporarily. Does a surfer who merely calls up a page of information onto a screen infringe copyright, even if the information is not downloaded for local processing or forwarded to any other person? Bizarre though such a suggestion may sound, it has been seriously debated—for example, during negotiations on a new international copyright treaty at the World Intellectual Property Organization in Geneva in 1996. A further aspect of the difficulty of defining copyright is that many countries have traditionally taken a more lenient position on private than public copying. But content producers now fear that such "fair use" exceptions, if transferred to the Internet, will blast a huge hole in their property rights and potential profits.

A third problem in attempting to enforce copyright on electronic products is maintaining national restrictions on distribution. At present, international copyright agreements allow books and music recordings to be sold differently in different countries. A publisher usually has exclusive right to distribute a book in a particular geographic area, say, Australia; another publisher, with rights to distribute the book in, say, Britain cannot freely ship extra copies into Australia. This protects large price differences for identical objects: in Sydney in spring 1997, a compact disc of a chart-topping album by the pop group U2 cost A$27.95 (about $22.30), while the same CD cost $13 in Los Angeles or Manhattan.[31]

But once an article or a piece of music becomes available on a Web site, anybody in the world can have access to it. In the future, books and music may have to be treated more like automobiles: a publisher who has bought publication rights from an author will be able to sell copies everywhere, sometimes in direct competition with another publisher issuing the same work in a different edition. This will be advantageous to authors and performers, whose interests are served by the widest possible dissemination of their work, as long as the payments due them can be enforced.

A fourth challenge to preservation of electronic copyright will be the increased difficulty of tracking down the source of illicit material. Here, as with other criminal on-line activity, electronic border-hopping exacerbates the problem. Pirated software may be put on the Internet by a person living in one country; it may be held in a computer in a second country; and it may be advertised through a computer in yet a third country. Some copyrighted material may deliberately be put on Web sites in countries willing to turn a blind eye to piracy. Many content producers fear the development of "piracy havens" around the world. The problem resembles that of policing pornography, fraud, and other crimes on-line: governments may attempt to hold ISPs legally responsible for knowingly making available illegal material.

Paying for copyright

At present, people generally pay for copyrighted material each time they use it: when a piece of music is played on the radio, a passage from a book is reproduced, or a film clip is incorporated into a new movie. Payment on such a basis becomes more difficult to enforce when digitized material can be reused, transformed, and distributed in many different ways. A few bars of some catchy tune, for example, may be incorporated into a movie, a video game, a commercial, a Web site, and so on. How can the use of such fragments be tracked, and how can the users be made to pay?

Establishing a payment system would be complicated, but it is not impossible. It requires, first, some way to identify the rights holder. In the case of books, an internationally recognized numbering system exists (the ISBN number, printed on the copyright page of this book) to identify the edition and physical volume (numbers differ from hardcover to paperback, for example, or for foreign editions). Other sorts of copyrighted material may have no such number: only since 1995 has an internationally recognized system been in place for numbering movies and television programs.

Second, an identifying sign, or watermark, would need to be attached not just to every copyrighted work as an entity but to fragments of that work: a few bars of music, say, or a few stills from a film. Several companies are working to develop such systems. Playboy, for example, is developing a digital signature to keep track of its pictures, which people are sometimes tempted to remove from Playboy's Web site to their own.

Third, a payment system would require that somebody keep track of fees due and send out the bills. This would require agreement on who pays and when. For example, should every user who visits an Internet site containing a particular picture pay the picture rights holder a tiny fee for each visit? Or should the site operator pay a larger fee, perhaps based on the number of users attracted? Should somebody consulting an item in an on-line guidebook be billed a fraction of a penny? Billing systems for micropayments are under development, but will be useful only if the cost of using them is less than the revenue they collect.

Redesigning copyright

The electronic media require new payment models that recognize the openness of new technologies yet allow copyright owners to make money from their material. As Ithiel de Sola Pool put it, "The question boils down to what users at a computer terminal will pay for."[32] Three routes for revising copyright regulations have been suggested.

First, as proposed by Esther Dyson, the American cyber-guru, content might be sold outright to advertisers or sponsors who will give it away happily in their bid to attract attention to their own physical products. Second, some theorists propose giving away the razor but charging for the blades: The Grateful Dead, a music group, long allowed people to record its concerts and made money from admission tickets instead, and plenty of consultants have made fortunes not from their books but from lucrative invitations to speak at conferences about them. Third, copyright might be replaced by a continuing relationship that provides service—dubbed by Ithiel de Sola Pool "serviceright"—with charges not for reproduction but for continuity of service, in the form of updating or maintaining the original material, for example. Not many people will bother to pirate a financial newsletter updated by the hour.

A workable system for protecting intellectual property in general and copyright in particular will have to be based on consent. At its heart should be the right to a fair return on an innovation rather than the right to exclude others from replicating it. The ease of replicating material in digital form will inevitably drive the law in this direction.

Living with Global Networks
. . . . .

The death of distance, in a world of national jurisdictions, poses intractable difficulties for regulators. National laws simply cannot provide the means for dealing with many of the regulatory problems that the Internet, in particular, creates. At the very least, governments will have to cooperate more than they now do on subjects as diverse as taxation, terrorism, and copyright.

But even with cooperation, many issues will resist effective electronic policing, except at a cost that rich democracies will be unwilling to carry. The death of distance acquires a second meaning here: it becomes impossible to put distance between criminals and victims or between terrorists and governments as can be done in the physical world.

The result will be an inevitable tradeoff will result between policing and easy access to information. Just as in the non-electronic world, where societies choose to accept some risk in exchange for a certain level of personal convenience and freedom, so it will be in the electronic. But in the electronic world, the balance will alter. The Internet will make government regulation harder, but self-regulation easier. Thus, although it will be easier to reproduce copyrighted material, it will also be easier for copyright owners to trace abuses. While access to pornography will be easier, citizens will have screening tools to protect themselves. Shady businesses may be easier to launch electronically, but with information readily available to everyone, customers will find it easier to investigate before they buy. Thus, while government will find it harder to protect citizens, people will need less government protection.

Regimes that refuse to recognize this point will also face a different tradeoff. They may be willing to accept more inconvenience and economic loss than are rich democracies, but the costs will rise over time. In a world where, for the great majority, knowledge flows freely, the penalties for restricting it will be impoverishment and marginalization.

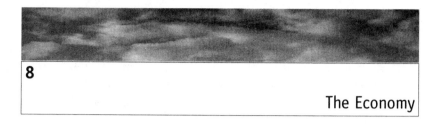

The Economy

Two tremendous changes are driving through the world economy: technological advance in computing and communications and globalization, caused both by the communications revolution and by the fall in barriers to trade and foreign investment. The death of distance brings the two together. As the world moves toward virtually limitless and almost free electronic communications capacity, trade and investment flows will transform patterns of economic activity around the world.

Many people fear these changes. They see technological advance as a threat to jobs. They worry that computers kill jobs. They fear the development of a disenfranchised underclass, deprived of job security and access to the complex technologies on which societies will increasingly depend. They fret that communities, already under pressure from the automobile and from crime, will be replaced by "virtual communities" in which people live their social lives electronically rather than on the streets.

People also fear the consequences of globalization. They see trade and foreign investment as threats. They worry that the industrializing economies, with their immense supply of low-paid workers, will deprive higher-paid workers in rich countries of their jobs. They already see money moving as never before: on average, some $1.3 trillion, or about one-third of a year's global exports, flashes around the world every day. How, in the face of such numbers, can countries control their economic destinies?

Put these two forces together, and you have a recipe for even greater terrors. In a world in which communication costs virtually nothing, any service can be sold on-line anywhere in the world. Jobs will move to the people. But what if the people live on the other side of the world?

· · · · ·

All of these fears are understandable. Change is always unsettling, especially change as rapid as we face today. Companies, economies, and individuals will be left with little time to adjust. Older people, in particular, learn more slowly than the young, and some of their skills, acquired over a lifetime, will become redundant. The rich countries already have populations older than ever before, and they will be older yet by the middle of the next century. So societies less demographically fitted for change than at any time in the past face faster and more pervasive change than ever before.

The changes will feed through different parts of the economy in different ways. Instead of the old divide between goods and services, we will see a new divide, between products requiring physical delivery and products that can be delivered on-line.

For many products, global markets will become more important than local or national ones. The change will not happen instantly: national regulatory barriers hinder international trade in services far more than they do trade in goods. Indeed that is the reason why the economic impact of the death of distance may be felt first in the United States, a large single market, but may ultimately be more profound in Europe and Asia, where more barriers frustrate trade. Ultimately, trade and foreign investment will become even more important, economies even more interdependent, than they are today.

This will have repercussions for jobs, for income distribution, and for productivity that will be traumatic for many people. But these changes will also be enriching, because their underlying effect will be to transform the availability of information and knowledge. All sorts of information—a business plan, a blockbuster movie, a blueprint, a battle order—can now be stored in a computer and sent across a telephone connection, making it infinitely diffusible. The barriers to the instant global spread of knowledge are falling away.

Knowledge is the basic building block of economic growth. It can also improve lives: a new technique for curing a sick child, for tilling a field, or for designing a car will spread rapidly around the world, spreading health and employment and wealth with it. It is important to remember that in the past, economic growth has always been built on technological advance. From the plough to the steam engine to the Big Mac, innovations have made society richer, not poorer. The communications revolution will do the same.

The New Divide

· · · · ·

One of the trends dominating economic development throughout the rich countries of the world for much of this century has been the decline in manufacturing as a proportion of the economy and the simultaneous rise of services. The United States provides the extreme example: services accounted for 57 percent of economic growth in the period 1972 to 1977, but 81 percent in period 1987 to 1992. But now, the familiar divide between manufacturing and services is becoming blurred. Many manufacturing techniques are being applied to the production of services and vice versa. These trends are being hastened by new communications technology.

The new convergence

In the past, manufacturing was most clearly differentiated from services by these characteristics. First, manufacturing produced a tangible object; services did not. Second, manufacturing operations could be at a distance from the final consumer; service operations could not. Third, manufactured goods could be mass-produced; services required individualized action.

One effect of the communications and computer revolutions will be to blur these distinctions. In the future, services will become more like manufactured goods and vice versa. If most of the value of a tangible product comes from intangible inputs, such as design and advertising, it becomes difficult to tell when the production of that object becomes a service, rather than manufacturing. If, on the other hand, life-insurance policies assembled on the other side of the world are sold to a mass market, does that insurance still constitute a service, rather than (in the word the life-insurance companies themselves increasingly use) a "manufactured" product. Thus, manufacturing output will be increasingly intangible; more services will be produced at a distance; and more services will be mass-produced.

Output is increasingly intangible
America's output, measured in tons, remains about as heavy as it was a century ago, even though real GDP, measured in value, is twenty times

greater. The main reason for this striking shift from material goods to intangibles, described in a speech in 1996 by Alan Greenspan, chairman of the Federal Reserve Board, can be identified as the rising proportion of total cost of the "knowledge" content of goods and services, relative to materials and energy. More and more value is added to products through design, styling, advertising, marketing, selling, consulting, and advising. Tangible goods have more and more knowledge embedded in them: a washing machine, for example, might include software that calculates the correct water temperature; a videocassette recorder might include software to control the timer or other functions. Intangible inputs now account for 70 percent of the value of an automobile.[1] By one estimate, the value of America's stock of intangible investment (R&D, education, and training) overtook its stock of physical capital (buildings and machines) during the 1980s.[2]

Services can be produced at a distance

Like manufactured goods, services can increasingly be produced at a distance from their final market. Financial services, entertainment, education, security monitoring, secretarial services, accountancy, and games can all be produced and sold at a distance from the ultimate consumer. One example is the Dyer Partnership, an accountancy firm in Hampshire in southern England. From there, the company provides the finance department of Windenergo, the Ukraine's leading manufacturer of wind turbines. Dyer handles all the work of administration and financial reporting, including preparing profit and loss accounts. The relationship operates almost entirely on-line.[3]

Such long-distance service provision is possible even in medicine, where some services no longer require that doctor and patient meet face to face. Once, a blood test meant a trip to a clinic and a disagreeable session as a nurse or a phlebotomist inserted a needle and filled one or more vials with blood. Now, two companies working jointly, one American and the other Japanese, have invented a gadget that, when simply placed on the skin, can read the broad chemical content of blood.[4] Since such information can then be stored in a computer and transmitted electronically to a doctor anywhere in the world, the two key aspects of a service—immediacy and the need for personal contact—may well vanish, even from aspects of medical care.

Services can be mass-produced

Low-cost communications allow the production of many services to become less personalized. Banking provides an example. In services, as in manufacturing, mass-production allows economies of scale. As long as retail banks needed clerks to sit all day at counters dealing with individual customers, a bank's growth was limited by the size and number of its branches, which constrained the number of customers who could be conveniently served. With banking transactions being handled by customers pushing buttons on an ATM, a telephone, or a computer keyboard, banks can grow much larger. Telephone banks can serve much larger markets, both in terms of geography and of customers per employee.

The new distinction

In the future, companies will sell goods and services to a much larger market, geographically, than ever before, and global marketing will be increasingly widespread. The new and really important distinction will be between processes and products that still require physical distribution and those truly intangible products that can be delivered on-line.

Global marketing

The ability to sell in global markets, the result of the rise of on-line marketing techniques such as the Internet and global toll-free telephone numbers, will encourage the growth of global niche services. Specialization has always, as Adam Smith noted long ago, gone hand in hand with market size: a village of a few hundred people may provide a market for a general store, but not for a specialist boutique.

This effect helps to explain why, for example, mail-order businesses often sell niche products (such as the smoked salmon purveyed by a small company in Scotland's Western Isles to the Vatican—until the Holy See ran up £30,000 of unpaid bills).[5] Now, such companies will be able to advertise to a world market. They will also be able to learn almost as much about distant niche markets as local companies can, thanks to techniques of data collection and mailing lists.

But where a product requires physical distribution, the impact of the death of distance will be lessened. For these goods, the more important

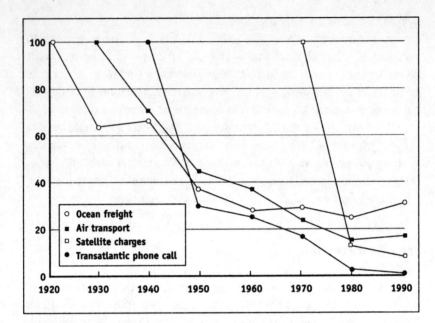

Figure 8-1   The Falling Costs of Transport and Communications
Note: The initial cost of each technology equals 100.
Source: World Bank, *World Development Report 1995*.

trends will be, first, this century's declining transport costs, both by sea and by air, but this may now be reaching its end, (See Figure 8-1.) and second, the rise in the value of physical products relative to their weight. The average weight of a real dollar's worth of American exports, for instance, has more than halved since 1970.[6]

The costs of distributing a physical product globally, relative to its value, will continue to decline. Put that together with the new opportunities for global marketing, and the growth of international trade, which has risen some fortyfold over the past half century, has clearly only just begun.

Trading intangibles
More dramatic than the effect of falling transport prices on tangible goods will be the effect of falling communications costs on those intangible processes and products that can be distributed on-line. There, dis-

tribution costs will fall much more quickly than have done those for physical goods—and much further.

Clearly, as experience in other markets shows, big changes in distribution costs can have enormous effects. The decline in the costs of ocean freight has revolutionized the human diet, for example, allowing ordinary people in the industrial world to enjoy a variety of food that could once be afforded only by the wealthy. The even more dramatic decline in air-transport costs has allowed ordinary tourists to visit places that were once the preserve of the rich. The rich may resent it, but for most people, the change has been liberating and enriching.

It took many years for these transitions to occur. The change in diet took half a century; the rise of mass tourism, now the world's largest industry, took about fifteen years. The fall in the costs of a telephone call or a satellite link has been even more substantial, and it is happening faster. But the full effect of this price shift will take time to become visible.

The effects will come first in trade between businesses, such as data processing and business software. From there they will spread first to retail services that can be sold on-line (software and financial services, for example), then to services that have not previously been widely traded but that can be delivered on-line.

Strikingly, some of the services that are the fastest-growing in many countries are hardly traded at all. Of America's fastest growing service industries—telecommunications, health and medicine, education, insurance, communications (including movies, music, and television), and travel—only the last three are widely sold abroad. U.S. exports of health and medical services, the country's largest single industry, are tiny compared with those of the aerospace (seventh largest) or electronics (tenth largest) industries. Clearly there is an opportunity here.

Rich countries have made big investments in education and medicine that may give them a comparative advantage over less developed countries. Already, Britain's Open University runs, by mail, telephone, and Internet, Europe's largest business school. Other universities will copy its success. In Alaska, a dedicated medical telephone network links local health workers in more than one hundred villages with doctors at a regional hospital. If such a network can be established within Alaska, why not between Alaska and, say, English-speaking countries in the Caribbean?

Suppose that we are on the verge of a new era, when all sorts of activities that we once thought of as local become available globally:

. . . . .

health care, education, financial and legal advice, and many others. One consequence would be replication of the democratizing effect of the falling costs of ocean freight and air transport. Services that could once be afforded only by the wealthy might, at last, be available to a mass market. Ordinary people would be, once again, liberated and enriched.

The Effect on Jobs

. . . .

The death of distance will have enormous implications for employment. The second half of this century has already seen a huge migration of jobs, driven by changing transport costs and falling trade barriers: between 1985 and 1995, foreign direct investment grew three times as fast as the world economy. Now, industries producing goods and services that can be sold on-line no longer need to be near their markets. Indeed, they can locate wherever they can find appropriate skills at the right price. The result will be a redistribution of employment within countries, a redistribution of employment between countries, and new fears of job losses.

Jobs within countries

Within countries, communications affect relations among different regions. Business has generally tended to concentrate in city centers, while people live in scattered areas. That may now change. Formerly remote areas of a country may have new opportunities to compete for business on equal terms with traditional population centers. This shift will have the greatest impact on developing countries, where recent migrations of population from country to city have been most pronounced.

Cities have traditionally been sited where communications and transportation links are good: by a port, on a river, at a railroad junction. In this century, the impact of the telephone initially allowed cities to become even denser: without it (and the elevator), businesses would have been much less willing to move into skyscrapers. Indeed, a writer in *Telephony*, a trade magazine, worked out way back in 1902 that if businesses in big American cities continued to rely on messengers, elevator wells in skyscrapers would have to be about twice their then size, making such buildings uneconomical.[7]

A large population and an agglomeration of businesses will retain one big advantage: it guarantees a large pool of labor, and so gives employers more choice, especially if they want to recruit workers with specialized skills. But this benefit has an offsetting cost: employees spend more time traveling to and from work. The rule of thumb in developing countries states that in cities of one million people, workers must travel an average of three miles to their jobs; in cities of five million, they travel seven miles.[8] Better communications networks, by making it easier for people to work near their homes, save time and resources and improve the quality of life.

At the same time, regions distant from a country's traditional centers gain new opportunities to compete. Call centers, for example, one of the new, footloose businesses in the world of inexpensive communications, can be sited more or less anywhere. These allow banks, mail-order companies, airlines, and travel and hotel firms to set up specialist switchboards to take many calls on a single number, which are then fed to a human operator who summons up information for or about customers while they wait to be connected. Some companies specialize in running such call centers for other businesses and have brought new work to regions such as America's upper midwest. In South Dakota, processing credit-card transactions over the telephone has become one of the most common forms of employment, and Omaha, Nebraska, is one of the largest telemarketing centers in the United States.

Developing countries stand to benefit most from the way good communications can tilt the balance of competitive advantage from cities to rural areas, because these countries are experiencing the fastest rates of urbanization, driven by the rural areas' lack of well-paid jobs and good educational, medical, and other facilities. This shift will bring many economic benefits.

Rural incomes will be larger. The telephone allows farmers to strike better deals for their crops. Before the introduction of telephone service to several villages and small towns in Sri Lanka, for example, farmers sold their produce to middlemen at prices that were 50 to 60 percent below the market price in Colombo, the capital city; the tele-

. . . . .

phone allowed them to keep up-to-date on prices, so that the farmers could demand—and receive—prices that averaged 80 to 90 percent of the Colombo price. Better-paid farmers have less incentive to move to the city.

Rural areas will gain more investment. In India, state officials in Maharashtra found that many companies' main reason for locating there was its telecommunications services, the quality of which thus seemed to determine which regions grew fastest.

Rural areas will experience a more equal relationship with the central government. In northern Canada, local leaders of an isolated native population found that their newly installed village telephones could be used to coordinate their negotiating positions before meeting with government or commercial agencies. The result was a more balanced and well-informed relationship.[9]

Good electronic communications, like good transport links, are vital tools for regional development. In Sweden, in the 1970s, the government encouraged telecommunications investment by firms and individuals in the country's sparsely populated northern region ("Norrland"), reducing their telephone charges by between 20 and 50 percent.

Most developing countries have not yet understood this point. Many do the reverse: they keep long-distance call charges high, even though these matter most to rural areas, to hold down the price of local calls, which mainly benefits city dwellers. In addition, many have a long history of underinvesting in communications and refusing to allow the private sector to do the job. In developing countries, even more than in the rich world, fair competition and freedom of investment are essential to reap the rewards of the death of distance.

Jobs between countries

Many different factors affect where companies locate, including wage costs, the skills of the work force, government regulations, and the political climate, but distribution costs are becoming steadily less important in this decision. In the future, time zones may become just as important a consideration for the location of some activities as geography has been. The world will move toward a three-shift world, with many services provided continuously around the clock by people whose countries are awake while customers elsewhere on the planet sleep. Financial markets already operate this way, with investment books successively passed from a dealing team in one time zone to a team in the next and so on around the clock and the globe.

Several countries—generally those with skilled labor forces, relatively low wage costs, and flexible work forces—have already spotted the opportunities this offers. Bangalore, for example, a city in southern India, has a flourishing software industry. Its exports, mainly to the United States, grew by 64 percent in 1995 to over $700 million. Some 104 of America's top Fortune 500 firms buy software services from firms in India, where programmers typically receive less than a quarter the wages that American programmers receive. One of the fastest growing parts of the industry, which is growing twice as fast as the American software sector, is "remote maintenance": repairing software for companies in other parts of the world, often taking advantage of the time zone difference to offer overnight service. Most of the clients of Infosys, one of India's largest software firms, with sales of $28 million in 1995, are in the United States. At the end of its working day, its American office e-mails Bangalore with customer problems, and Bangalore technicians solve them while the Americans sleep.[10]

Ireland, pitching to be Europe's main international call center, offers another example: it is building on its lavish supply of college-graduate labor, low wage costs for white-collar jobs, relaxed labor laws, geographical position between Europe and the United States, and cultural links with the United States. Telecom Ireland has an intercontinental service that allows companies to link their European and American call-centers, thus taking advantage of the time differences between them. Among the companies Ireland has attracted are Gateway 2000, an American company specializing in direct sales of PCs, which employs fourteen hundred people on the edge of Dublin, including some three hundred non-Irish Europeans; Kao Infosystems (Ireland), a Japanese-owned firm that handles calls for Microsoft, Hewlett-Packard, and Virgin, with Dublin agents handling calls in thirteen languages, including Hebrew and Afrikaans; Oracle, an American software giant, whose ninety sales agents sell directly from Dublin to the rest of Europe and which plans to expand sales to Africa and the Middle East;

and United Parcel Service, an American courier, which is setting up a single center in Ireland to handle its German, French, Dutch, Belgian, Swiss, and Nordic customer services.[11]

Another example: on North Uist, an island thirty miles off the northern coast of Scotland, thirty people make their livings by producing abstracts of articles from American newspapers such as the *Salt Lake Tribune* and the *Denver Post* for an on-line database run by Information Access, a Californian company. The company's other database workers are in the Philippines.[12]

In the future, good communications may allow the growth of more companies like EMS Control Systems in Perth, in Western Australia, which monitors the air-conditioning, lighting, elevators, and security in office blocks in Singapore, Malaysia, Sri Lanka, Indonesia, and Taiwan. But, for the moment, the pioneers of the footloose company are often either in marketing and customer services or in data processing. Call centers, for example, often see time zone as an important locational criterion. Matrixx Marketing, part of the Cincinnati Bell Group, has seventeen centers around the world, many taking calls generated in response to television advertisements. Such calls tend to come within the first few minutes after an advertisement appears. So at peak periods, other centers provide extra capacity. The company's center in Salt Lake City, for example, takes over when its office in Newcastle-on-Tyne, in the north of England, is fully stretched.[13]

Airlines routinely outsource their back-office processing halfway around the world. Examples include British Airways, which decided in 1997 to transfer much of its passenger revenue accounting to a subsidiary in India, and Swissair, which puts its ticket accounting and computer-entry queries through Bombay.

Insurance companies increasingly process claims at a distance. Caribbean Data Services, set up by American Airlines, is the largest employer in Barbados, where as many people now work in information technology as in growing sugar cane; it processes claim forms for several U.S. insurance companies.

"In a world of information technology, it doesn't matter where you are," says Peter Scott, who runs Oracle's Dublin-based European direct-marketing operation.[14] Today, companies like his still need to have their staffs physically in one or a few places and therefore need to bring the workers to the jobs. That means recruiting people who understand something of the distant markets with which they deal: people who can cope with different languages, currencies, spelling— and humor.

In the future, though, the locational freedom that Oracle enjoys will extend to its staff. Once workers anywhere can communicate with their employer as inexpensively as their employer communicates with the market, jobs will truly be able to go to the workers. If people want to work from their garden shed rather than from a downtown office, they will be able to choose to do so.

Worries about jobs

Many in the rich world fear that production and jobs will shift on a terrifying scale from the rich world to the poor. People in white-collar jobs in rich countries fear that their jobs will vanish down a telephone line. They see themselves exposed to the competitive pressures that have already squeezed manufacturing workers.

Certainly, some rich-country activities will migrate to poorer countries; and the most footloose industries will move on from the richer industrializing countries to the poorer ones. Basic data processing, for instance, is already shifting to China and Vietnam. But, where labor markets are reasonably flexible, new jobs will appear. Charting employment growth against investment in information technology yields reassuring results. (See Figure 8-2.)

When old jobs go, everybody notices. When new industries spring up, they often do so unannounced. Many new jobs will emerge through the export opportunities opened up by low-cost communications. As John Heilemann, an American media journalist, puts it:

> More Americans make computers than cars; more make semiconductors than construction machinery; more work in data processing than oil refining. Since 1990, U.S. firms have been spending more on computers and communications gear than on all other capital equipment combined. Software is the country's fastest growing industry. World trade in information-related goods and services is growing five times faster than in natural resources.[15]

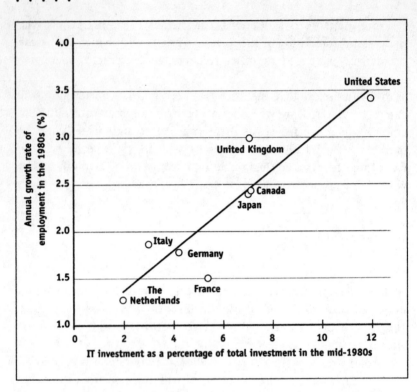

Figure 8-2   Information Technology Investment and Employment Growth
Source: *Information Technology Outlook 1995,* ©OECD, 1995, Paris. Reproduced by permission of the OECD.

Heilemann could have added that, since 1990, America's movie industry has created more jobs than its automobile manufacturers, pharmaceutical firms, and hotels combined; that the computer-software industry employs ten million people around the world; or that call centers, an industry that barely existed a decade ago, now employ four million people in the United States and over a quarter of a million in Europe.

All these jobs require skills such as articulateness, courtesy, creativity, accuracy, and resourcefulness, which schools and universities need to help their students develop. These skills, not specific to gender, age, or race, will result in more competitive work places. Opportunities will be greater, but more people will also be able to compete for them on equal terms.

If people are flexible, and if government regulations do not inhibit entrepreneurs, new jobs will arise, often in industries that did not exist a decade or two previously. But, just as the pattern of employment alters, so will the pattern of pay.

The Effect on Incomes

. . . . .

The death of distance will affect incomes in ways that at first seem paradoxical. Between countries, incomes will become more equal; within countries, incomes will become more unequal. Everywhere, the premium on skill, creativity, and intelligence will rise.

Income between countries

Wages, especially in industries producing goods traded internationally, tend to track productivity. In 1990, manufacturing wages in Malaysia were only 15 percent of those in the United States—but manufacturing productivity in Malaysia, too, was only 15 percent of that in the United States.[16] As productivity in developing countries rises to rich-world levels, their wages tend to do the same. Asian countries such as Hong Kong and Singapore now sell their skills and state-of-the-art investment, rather than their low wages: their economies have shifted toward high wages and high productivity.

One effect of the death of distance will be to increase the share of every economy that is traded. Another will be to make companies more footloose—more willing, like Oracle's Mr. Scott, to locate wherever the best bargain of skills and productivity can be had. Wages will be calibrated more precisely, in more industries, with a world standard of productivity.

The discount now imposed by the economy for lack of skills will be driven down even further. In the United States, since 1980, the lowest-paid 10 percent of men have seen a decline of almost 20 percent in their real wages; the top 10 percent have enjoyed a real pay rise of roughly 10 percent. Pay inequality has also increased in Australia, Britain, Canada, and New Zealand. In France and Germany, inequality has diminished—but unemployment has shot up. If these countries dismantled the barriers and Social Security benefits that discourage the least skilled from seeking work, they might well see a decline in the real pay of the unskilled.

Income within countries

While the relative pay of the skilled has been rising, something much more dramatic has been happening in a few occupations. Earnings have become less equal, not only between occupations but also within them. Chief executives, bond dealers (even honest ones), film stars, and sports stars have all seen an extraordinary dispersion of income, so that a few earn astronomical sums, compared with the majority. This phenomenon, described in a book by two American economists as the "winner-take-all society," relates directly to the death of distance.[17]

Three simultaneous changes contribute to this effect. First, many more products cost little or nothing extra to reproduce millions of times, such as Windows95, a hit record, or a movie video such as *The Lion King*. It costs no more for twenty million people to watch the Oprah Winfrey show than for a dozen to do so. Second, inexpensive, fast communications enormously widen the market for many products. Buyers and sellers can now make contact with each other over vast geographical distances and discover the global best, rather than the local best. A singer who would once have charmed a city now has a potential world market; a heart surgeon who might once have operated only on the citizens of a particular state can now be a world celebrity.

Thus, as electronic reproduction slashes costs of making perfect copies, electronic distribution puts star performances in front of a much larger market than ever before. A huge increase in demand for a product that was once in finite supply can now be satisfied at little extra cost. Result: the rise of the superstar. A few top performers, in sports, music, movies, and television, make mind-boggling amounts of money. But the nearly-as-good make far less. People want to listen to Madonna or Michael Jackson, not to a rival who might happily accept a tenth of their income and sound, to the uninitiated, just about as good.

These effects are compounded by the third change, what might be called the "celebritization" of the economy. In a world of infinite information, the simplest way for a company to draw attention to a product is to pay a global celebrity to endorse it. The few names with global resonance—Madonna, Michael Jackson, or Britain's weight-watching former Duchess of York, say—attain far greater worth in this game than the merely fairly successful sports or entertainment person with a modest national or local following. Wally Masur, a top Australian tennis player, was a semifinalist in the 1993 U.S. Open. At no time in his career did manufacturers employ him to endorse tennis shoes or rackets[18]—being a semifinalist is simply not good enough.

The superstar phenomenon will spread to other jobs. The market will pay lawyers, chief executives, bankers, and doctors not according to their absolute performance but according to their performance relative to others in the same field. In those jobs offering the greatest rewards for being or having the best, the best will earn fortunes.

The Effect on Productivity

. . . . .

In the two centuries since the industrial revolution, the driving force behind the global rise in wealth has been increased productivity. New mechanical and electrical technologies, together with the development of the factory and of mass-production techniques, have allowed people to produce vastly more per head than ever before. Will the revolution in information technology driving the death of distance have an equally dramatic impact on productivity in the next century?

The impact of the communications and information-technology revolution on productivity has up to now been hard to discern—indeed, economists have been led to speak of a "productivity paradox": more investment, but lower productivity. Rises in productivity, however, are almost certainly occurring—and will continue to occur, although in ways that are especially difficult to measure.

The productivity paradox

The rich countries have invested massively in information technology in the final quarter of this century: in the United States, it jumped from

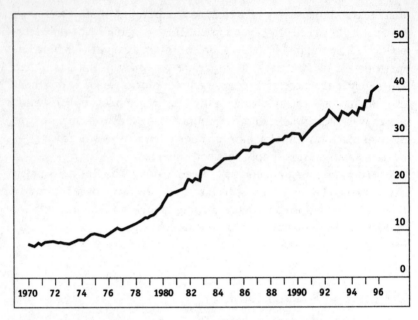

Figure 8-3   Corporate Investment in Information Technology
IT as a percentage of U.S. firms' total investment in equipment, 1970–1996.
Source: Datastream, ©*The Economist*, London (September 28, 1996).

7 percent of corporate investment in 1970 to more than 40 percent today.[19] (See Figure 8-3.) Capital spending on computers and communications now exceeds investment in traditional machinery. Total spending on computers alone in 1996 accounted for 3 percent of America's GDP, almost as large a share as spending on autos and light trucks, which accounted for 3.5 percent.[20]

In theory, all this expenditure should have increased productivity and economic growth. Instead, productivity increases in the rich world have been tailing off. In the seven main OECD countries, labor productivity, or the amount of output each worker produces, grew three times faster between 1960 and 1973 than it has done subsequently. Most ominously, productivity growth in the United States, where spending on information-technology investments has been highest, has been lowest of all.

Nearly all the economic growth since 1973 in the seven big OECD nations has been the result of more labor (women joining the labor force, for instance) or more capital (buying more machines). Hardly any has been caused by technological change. In the words of Robert Solow, an American economist and Nobel Prize winner: "You can see computers everywhere but in the productivity statistics."[21]

So where has all that investment in information technology gone? One possible explanation is that, lavish though the investment has been, it remains small compared with the total capital stock. Certainly computers account for a mere 2 percent of America's capital stock, tiny compared with the railways at their peak in the late nineteenth century, for example, when they were 12 percent of America's capital. But add in all the other paraphernalia of information technology, including telecommunications and software, and the total approaches the railways' 12 percent.[22] The paradox remains.

Three other possible explanations have been proposed. First, plenty of company spending on information technology has, in some sense, been wasted. Lots of people have computers on their desks with vastly more power than they will ever use. Indeed, it is far easier to waste time with a computer, especially one connected to the Internet, than it ever was with a typewriter. Companies have installed systems only to find that they do not do what they wanted done, or that the new system is incompatible with the system of another business with which they want to merge, or that it will crash at the turn of the millennium unless expensively reprogrammed. Whole new departments have been set up to manage corporate investments in communications and computer networks. No wonder many managers see information technologies as a bottomless pit, into which immense sums of money vanish with little trace.

Some economists argue that investment in computers and communications genuinely increases productivity less than some earlier inventions. Paul Krugman, of Massachusetts Institute of Technology, points to the much greater savings in people's time that came from mass air transport or the supermarket. "Computerized ticketing is a great thing, but a cross-country flight still takes five hours; bar codes and laser scanners are nifty, but a shopper still has to queue at the checkout."[23]

Second, economies can easily take a couple of generations to learn how to get the most from a new technology. The dynamo, which opened the way for the commercial use of electric power, was introduced in the early 1880s.[24] Yet, even at the end of the century, electricity accounted for less than 5 percent of the power used in American

. . . . .

manufacturing; even in 1919, it accounted for only 50 percent. At that stage, thanks to Henry Ford, companies began to realize that electricity not only served as a source of power, but it also allowed machines to be used quite differently. Machines that had previously been placed next to water wheels or steam engines could, because of electricity, be placed along a production line, thus maximizing the efficiency of the work flow.

It took forty years, from 1890 to 1930, for the cost of electricity to decline by 65 percent. Computing and transmission costs are falling faster, which will drive their exploitation faster. But technologies may still need to reach a critical adoption level before their effects are significant.

In the case of electricity, the big gains in productivity started to appear when it accounted for 50 percent of power. In 1984, computers were used, in some form or other, by a quarter of American workers; today, that figure hovers around 50 percent. But the IBM PC was not introduced until 1982; and the spread of the networked computer only began after that. With hindsight, those may well appear to be the key landmarks. It may also be that greater opportunities for productivity gains will result from the impact of the networked computer on interactions such as data gathering and communication than from the power of the isolated computer to process and store information.[25] If so, the surge in productivity may be beginning. If electricity is a good model for the future, the surge will run through the first quarter of the next century, bringing a long period of economic growth more rapid than anything seen in the rich world since the early 1970s.

The third explanation for lack of productivity increases may be that productivity may, indeed, be improving, but in ways difficult to measure. Productivity statistics tend to measure change in the production of goods. They do not take full account of new goods that appear nor of the quality of goods. The value of many consumer durables—think of the PC—now lies in the software programmed into it rather than the hardware itself.

Productivity growth in services is particularly difficult to measure. It appears to have been slow, even though service industries have been spending the most on information technology. In manufacturing, investment in information technology has been running at about 3 percent of output. In air transport, telecommunications, retailing, health care, banking, and insurance, investment in information technology has averaged almost 6 percent of output.[26] The explanation that productivity increases are hidden may be nearest the mark.

Measuring productivity in services

Measuring changes in the productivity of services involves solving two problems: measuring output and measuring quality improvements.

Changes in output

With physical production, output is easy to measure. So with agriculture, you can count the turnips each farmer grows; with car manufacturers, you can count the wheels that drive out of each plant. With services, output measurement is more difficult. Statisticians typically assume that the output of a nurse or a school teacher or a caregiver rises in line with the number of hours they work. Measuring electronic-mail messages or inquiries handled over the telephone or the effectiveness of television advertising is more difficult still. Neither physical production nor time spent is a guide.

Improvements in quality

Many investments show up in better quality rather than in larger quantity. Digital television, for example, will bring viewers a sharper picture and clearer sound. Even with physical goods, improvements in quality present problems for statisticians. But while quality in an automobile might be measured partly in terms of reliability, with a television set, quality is in the eye of the couch potato.

A result of these two characteristics—the immeasurability of output and the subjectivity of quality—is that some increases in productivity may actually show up as declines in output. If a networked computer allows a nurse to select and call in patients for a life-saving test, such as a cervical smear, at more appropriate intervals, the nurses may do fewer tests and yet find a greater number of people needing treatment. Their measured output may appear to decline, but the quality of service has improved. Many other examples would show the same effect. Automatic teller machines allow customers to draw money around the clock; credit cards make it possible to buy all sorts of goods and services over the telephone or on the Internet. These improvements in convenience not only entirely bypass national accounts but the number of checks processed will appear to decline, as well, and might be measured as a fall in bank industry productivity.

The largest service providers of all are governments, which presents a third measurement problem. Many productivity increases may eventually appear in delivery of state services, such as welfare, tax collecting, and education, but as these services are usually given away rather than sold in the market, their value is even harder to establish than that of the television industry's output. In fact, productivity improvements in government services may be genuinely slow to take place. Governments are notoriously slow to adopt new technologies: adoption requires investment, and public spending is being universally squeezed; slowed spending will probably slow the pace at which governments learn to make the most of new communications and computers.

In general, though, information technology is almost certainly raising the productivity of the service industries, but in ways that are difficult to assess. A hidden boom is taking place, improving the quality of all sorts of services in ways that consumers dimly notice but that the published figures ignore. These improvements enrich us, not by giving us more affordable material possessions but in a more subtle and fundamental way. Improvements in services such as health care and entertainment will have more positive effects on our quality of life than improvements in material goods are henceforth likely to have.

Growth and the Knowledge Revolution
. . . . .

The death of distance will transform the availability of information and knowledge, the fundamentals of economic growth. New technologies, new products, and new ideas will spread faster than ever before. Markets will work better. The market economy, now adopted in almost every country on earth, will allocate resources more effectively on the basis of ever better information.

All these changes will accelerate the adoption of innovations. Developing and launching new products will be less expensive, and potential

customers and investors will be easier to find. The links between innovators and their customers will grow shorter.

To see what lies ahead for many industries, consider the new companies springing up around the Internet, scrambling to refine and extend
this new form of communications. Internet companies form around a
promising idea; they both cooperate and compete with rivals; they
boom or die remarkably quickly; they flourish on a plentiful supply of
clever people and of venture capital. In the knowledge economy of the
future, this will be the template for many industries.

Essential to smoothly working markets are prompt price information
and low inflation. The death of distance will help achieve both. Many
more companies and producers, like the Sri Lankan farmers mentioned
above, will have access to accurate price information. In addition, many
services that could once profit from effective local monopolies will be
competitively and globally traded, as is happening most obviously in
telecommunications. Margins will be squeezed, wiping out excessive
profits and helping to curb inflation.

But for communications to achieve the maximum effect on the world
economy, two things are essential. First, the world market in communications—not just in hardware, but in communications services—
must be liberalized. That means removing regulatory constraints that
now discourage or distort investment. In too many countries—especially in the developing world—telecommunications are still protected,
monopolized, and state-owned. As a result, those countries have little
chance of enjoying the economic benefits that the death of distance
promises. Their telephone systems in particular will be inadequate,
unreliable, and badly maintained.

At the beginning of 1997, some seventy countries took a big step in
the right direction by agreeing at the World Trade Organization to open
their markets to unfettered domestic and international competition.
The three largest telecommunications markets—the United States,
Japan, and much of the European Union—promised to start the
process in 1998, and fifteen other countries, including South Korea and
Mexico, agreed to join them. Others plan to follow in subsequent years.

If competition blossoms, its positive effects will be tremendous. But
achieving competition in communications will be difficult. Encourag-
ingly, nearly forty countries agreed at the WTO to allow some measure
of foreign investment in telecommunications starting in 1998. But,

unless governments back competition with conviction, it will come slowly and patchily.

The second development essential before communications can achieve their maximum potential in the world economy is that world trade in services in general must be liberalized. The argument for free trade in services is just as persuasive as the case for free trade in goods: countries can make themselves richer by specializing in activities in which they have a comparative advantage, whether it be growing coffee or producing blockbuster movies.

But, while trade in goods has been extensively liberalized in the second half of this century, all sorts of barriers stand in the way of trade in services. It is much harder to buy, say, health insurance from a foreign insurance company than to buy a foreign automobile or videocassette. And, while many governments pay foreign construction companies to build roads or power plants, they are much less likely to pay for their citizens to consult foreign doctors or to learn from foreign teachers.

Liberalizing trade in services will not be easy. Often not customs duties or tariffs but domestic regulations and rules about qualifications and the like create barriers to foreign services. Think of the obstacles to trade in domestic banking, medicine, law, legal services, and so forth. Without ways to harmonize such regulations, or to dismantle them, international trade in some services will be held back. The world will be the poorer as a result.

New opportunities lie ahead to free trade in services. In 2000, a new round of talks will begin at the WTO. The WTO's predecessor body, the General Agreement on Tariffs and Trade, required more than forty years to liberate trade in goods. The economic opportunities created by the death of distance mean that the liberation of trade in services must and can happen faster.

In fact the most powerful force for liberation may be communications itself. The more people who can buy services on-line, the harder it will be to impose barriers to prevent them from doing so. Good communications will therefore drive liberalization, improve the quality of services all over the world, and extend consumer choice.

Society, Culture, and the Individual

The next quarter century will see the fastest technological change the world has ever known. How will that affect our lives? In general, as Arthur C. Clarke once pointed out, people exaggerate the short-run impacts of technological change and underestimate the long-run impacts. Really big technological changes permeate our homes, our personal relations, our daily habits, the way we think and speak. Consider the links between the automobile and crime, or between electricity and the skyscraper, or between television and social life. Each of these technological advances had consequences that nobody could have foreseen when they were new. The revolution in communications will have consequences that are just as pervasive, intimate, and surprising.

Many people fear that future. They see a society of isolated people, stuck indoors, glued to a screen, losing the taste for real human contact and experience. They worry about the exclusion of the poor, the old, and those too inept to learn how to plug in their modems. They imagine a new class of technological have-nots, as socially deprived as anybody without an automobile or a driving permit is often presumed to be. They see an Orwellian world of lost privacy. In countries other than the United States, people fear a future in which everybody speaks English and thinks like an American, with cultural diversity engulfed in a tidal wave of crass Hollywood values.

In fact, the main impact of the death of distance will be to make communication and access to information in all its forms more convenient. On balance, that will surely be good for societies everywhere, although the nature of the effect will depend on why people communicate and what knowledge they choose to acquire and how they use it. Like the automobile, telecommunications technology will be a tool that can be used for

. . . . .

bad purposes as well as good, but it should, overall, make people's lives easier and richer. More communication is nearly always better than less.

In broad terms, the societies of the rich world may be altered in four main ways. The role of the home will change. The home will reacquire functions that it has lost over the past century. It will become not just a workshop, but a place where people receive more of their education, training, and health care. New kinds of community will develop, bonded electronically across distance, sharing work, domestic interests, and cultural backgrounds. Both the use of the English language and the strength of local cultures will be enhanced. English will be the standard global tongue, but many regional languages will revive, as well. And, finally, the young, and especially the intelligent and well-educated young, will be the winners in the new electronic world, and young countries will gain an advantage over older ones.

So society will be different: communities will take different form, and people will take for granted things that now seem strange. But the biggest changes will take longest. The death of distance will shape the world of the mid twenty-first century much more profoundly than the world of the next decade.

Work and Home

. . . . .

Amid the speculations is one certainty: the falling price of communications will affect where people work and live. The old demarcation between work and home will evaporate. In its place will be a shift in the location of work, a new role for cities, a new role for the home, and reshaped communities. Of these, the third may be the most profound. The home will once again become, as it was until the Industrial Revolution, the center for many aspects of human life rather than a dormitory and a place to spend the weekend.

The location of work

In the past century, people have tended to live farther and farther from their place of employment. Instead of living on the farm or over the shop, people have moved from city centers to suburbs and from suburbs to commuter towns. In some cities, almost all the decline in working hours in this century has been gobbled up in increased commuting time.

One of the main reasons for this trend has been the concentration of employment in large units. Through the first half of the century, the size of the factory, and then of the office, tended to grow. With factories in most rich countries, this trend had begun to reverse by the 1970s, mainly as a result of the rise in productivity.

Now, the nature of working life is starting to change again. A primary shift in the location of work will affect people in three main categories: routine teleworking jobs, vehicle-based jobs, and jobs involving top-level personal contact.

Routine teleworking jobs

Initially, teleworking jobs will be concentrated in large units. A bank no longer needs as many small branches once people can do more of their banking by telephone and by other electronic means; a travel agent no longer needs as many people in local offices. Such activities can be handled by a single call center, which does not need to be located in a particular city. The activity becomes centralized, but the place of work can be decentralized.

Some teleworkers will work from home. Various experiments have found a number of benefits in working from home. Foremost is more job satisfaction and lower stress. A trial by Northern Telecom in 1994 found a 30 percent increase in productivity when five hundred employees worked from home at least three days per week. Another benefit is increased flexibility in working hours. A one-year experiment conducted in the far north of Scotland by Britain's BT, the country's primary telephone company, found that home-based directory-assistance operators, connected to a supervisor and the customer by a telephone line and a computer link, were more reliable.[1] Environmental benefits also accrue from home-based workers. Results from a trial sponsored by AT&T in Phoenix, Arizona, in the early 1990s, suggested that if 1 percent of all workers in private and public organizations of one hundred employees or more telecommuted one day each week, fuel consumption would be cut by almost 500,000 gallons per year.[2]

Electronic home-working has, of course, some significant drawbacks, as well. Connection to a high-capacity fiber-optic network constitutes an expensive investment. If employers carry the cost, they will do so initially only for the workers with the most valuable and irreplaceable output. Maintenance costs will also be high. When connections or equipment break down, repairs may be easier in a central office than for workers scattered around the country. As long as their equipment is out of action, so, too, are the teleworkers.

"Telecottages" may offer a solution to problems of cost and maintenance: these offices will spring up in many small towns, allowing a group of people to share high-quality communications facilities in one building. This will have the added benefit of overcoming the sense of isolation that some home workers feel.

Vehicle-based jobs

Mobile communications will allow the growth of a whole category of workers—many of them engaged in various kinds of maintenance work or in delivery (of people as well as goods)—whose days will be structured by controllers using a mixture of satellite positioning equipment to track their teams, computers to calculate the most efficient routes, and mobile telephones with messaging systems to maintain contact between dispatchers and workers.

Safeway, for example, Britain's third-largest supermarket chain (no longer related to the U.S. chain of the same name), uses satellite tracking to follow the movements of its six hundred trucks, reducing delays and streamlining delivery routes. An increasing number of workers will thus, as it were, telework from a vehicle rather than from home.

Jobs involving top-level personal contact

Many senior employees will find themselves doing their work almost anywhere but in the office. The number of "portfolio" workers—nonexecutive directors, consultants, and one-person agencies who combine several senior-level freelance jobs under contracts with different companies—will grow. Such people will not have an extensive corporate structure behind them. They will keep their own complicated diaries, sift their own mail (as, electronically, Bill Gates of Microsoft is said to do), and handle their own intricate travel arrangements. They will work partly out of their homes and partly out of cabs or airport departure lounges.

Airports and airplanes will be, increasingly, temporary offices for senior people, who, ironically, will spend more time on the road as a result of better communications, even as the same communications decrease the time their staffs spend commuting. To hold together businesses with global scope, which will often have production and distribution spread among several countries requires not only an efficient electronic network but an enormous amount of travel by managers and suppliers. For these people, living near a good airport may be even more important than living near the head office. They will need and want portable offices: a multipurpose mobile telephone that will receive messages, send faxes, contact the office database, and tap into the Internet. For them, the most advanced communications will be godsends, as long as the equipment and services deliver what they promise and work reliably.

These workers may also operate from restaurants, at conferences, in golf clubs. All these venues provide a much more appropriate backdrop than the office for the face-to-face contacts that are central to many service industries. A senior partner in an advertising agency cannot deal with clients entirely by telephone any more than can a high-powered lawyer or a Hollywood agent. These personal contacts require a structured informality, a blend of the personal and the professional. The telephone and electronic mail may reinforce basic contacts, but they cannot substitute for face-to-face encounters.

What of the office? It will become the place for the social aspects of work, such as networking, lunch, and catching up on office gossip. The familiar roles of home and office will thus be inverted. The provision of a good office canteen, sports facilities, and plenty of meeting rooms will become more important; the individual office, with a large desk and lots of storage space, less so. The office will become a "club," a transformation spotted by Charles Handy, a British management guru.[3] The office-as-club will foster a sense of fellowship (or corporate culture) among employees, stimulating conversation and belonging; it will be the main site at which companies stimulate workers and mold them into teams; and it will provide a location for training, communicating, and brainstorming.

In the future, then, much more work will be done outside offices. Some will take place in transit, some in the home. That will have long-

· · · · ·

term consequences for city centers, where in the twentieth century work has increasingly been located.

The future of cities

In half a century's time, it may well seem extraordinary that millions of people once trooped from one building (their home) to another (their office) each morning, only to reverse the procedure each evening. Commuting requires a transport network built to cope with these two peak daily migrations. Roads must accommodate the weight of rush-hour traffic, and commuter railways and buses must carry the mass of peak-load passengers. Commuting wastes time and building capacity. One building—the home—often stands empty all day; another—the office, often in the most expensive part of town—usually stands empty all night. All this may strike our grandchildren as bizarre.

The dispersion of work will alter the nature of cities and towns, reaffirming their importance as places where people live and as centers for entertainment and culture. More people will return to living in city centers. Cities with attractive architecture and plenty of shops and cafes providing lively street life will delight those who relish the bustle and the anonymity of living in a crowd. Cities will become even more attractive as they become safer, thanks to wider use of electronic surveillance. Cities will also thrive as centers of entertainment and culture: places to which people travel for a stay in a hotel, a museum or gallery visit, a restaurant meal, or to hear a concert or band. Many kinds of work will require entertainment: managers will increasingly want to take valued customers to a theater, a restaurant, a club. So some of these delights will be paid for, directly or indirectly, by companies.

Suburbs and rural towns will benefit from the telecommunications revolution in other ways. Many people may once again "live over the shop" and work in their communities, rather than commuting. The benefits will include more revenue for local stores and other services, as workers stay close to their hometowns during the weekday; more opportunities for delivery services, since ordering over the Internet or telephone becomes easier if someone is at home when the product reaches the doorstep; and less local crime, because homes will be occupied and streets will be busy during the day, making them safer for everyone.

All these changes will have a darker side. The fragmentation of the large employer has made many workers feel less secure. More and more people, employed on short-term contracts or as freelancers, will be only as good as their last job and under constant pressure to find the next assignment before the current one finishes. The blending of leisure and work may well mean, in practice, that work increasingly intrudes into leisure: it makes the more forceful demands. Besides, working at home, however sophisticated the electronic links, lacks the companionship of working alongside other human beings. And, of course, not every downtown office complex, if deprived of office tenants, will be quickly transformed into a glittering entertainment center. But on balance, the direction of change may be to restore communities, to improve the quality of cities, and to give people more control over their working lives.

The future of homes

Among the most striking of all changes brought about by the death of distance will be a shift in the role of the home. In the future, people will not only entertain, relax, and sleep at home; they will increasingly find there a range of services, from health care and education to investment and employment. From their homes, people will be able to draw electronic money on a plastic card, study any subject from astronomy to zoology, seek legal advice, participate in a political debate, or bid in an auction.

Other home-based services may come to include monitoring (and sometimes repair) of domestic machinery without a home visit; monitoring of the housebound sick or elderly; and a level of security monitoring now available only to offices and the wealthy. Some of these services will become possible once homes have a permanently open telephone line or a computer with a camera attached permanently online. Such an arrangement would enable a family on one side of the world, for example, to keep an open line to a grandmother on the other side, or for caregivers (who may work around the clock, with three shifts in three different time zones) to monitor elderly people in their homes as much as the old folk need and want. Labor-intensive monitoring services will probably be provided by countries with cheap labor, with a hotline back to the neighborhood so that on-the-spot help can be available quickly.

Such changes will alter the design of the home. Architects have not yet caught up with its mutation from a place where people consume (meals, entertainment, and so on) back into a place where people also produce (today's homes can now be equipped with as much computer power as a large factory might have had in the 1970s). They now need to find a comfortable way to accommodate the home office.

For many people, the home office remains an unsolved problem: it doubles as a spare bedroom or crouches uneasily in a corner of the living room. Until the home office finds a permanent resting place, home design will need to be adaptable, so that flexible use of space can match flexible work patterns. The loft, with its vast open space, may be a more useful prototype for the twenty-first century home than the houses of the 1950s and 1960s, with their small, purpose-designed rooms.

New Communities

• • • • •

The concept of cyberspace—a computer-generated, three-dimensional world in which people live in virtual reality—was conceived by William Gibson, a science-fiction writer, in a book called *Neuromancer*, published in 1984.[4] The disembodied world he presciently portrayed symbolizes for many people the danger that electronic communications will be isolating and inhuman, as people find a social life in the chat rooms of the Internet.

But it is perverse to imagine that a technology that makes it easier to communicate should simultaneously reduce human contact. New forms of communication may change the nature of that contact—as the telephone call has replaced the gossip in the village street—but, more probably, it will increase the variety of ways people can and do communicate. At times, people prefer the privacy of a telephone call to a gossip in full public view.

The main impact of better and less expensive communications will be to create new ways to socialize and to build communities of interest, independent of geography. Both will enrich people's lives and mitigate the effects of separation that go with the increase in international migration, overseas employment, and business travel.

The future of socializing

Communications will allow a new kind of social life. Already by far the most popular use of the Internet has been electronic mail, and social calling constitutes the fastest growing area of international telephone traffic. Some people will want to talk to friends, some to strangers. Talking to friends will become the telephone's main use. Indeed, Britain's BT reckons that 55 percent of the calls it carries out of Britain are made from homes—some of them, perhaps, encouraged by its advertising slogan: "It's good to talk." Such social calls tend to last for twenty to thirty minutes—far longer than the brisk two to three minutes typical of a business call—and to be made for the most part by women. One study found that the predominant overseas caller from Australia was a young professional woman, born in another country, making long social calls to friends back home.[5]

International migration, business travel, and tourism all increase the number of people separated from their friends by distance. More telephone calls go from Germany to Turkey, for instance, than from Germany to the United States. While the United States is a much more important trading partner for Germany than is Turkey, Turkey has been a much bigger source of immigrants. Electronic communications allow friendships and families to survive separation.

Communicating socially with strangers will be a long-term use for the Internet. Unlike the telephone, the Internet creates communication with minimal personal contact: messages are merely typed on the screen, with no need even to disclose one's voice, let alone personal appearance, age, or gender. And the Internet offers scope for one-to-many communication (through bulletin boards and "chat" rooms) as well as one-to-one.

Prodigy, an early on-line service, discovered this with surprise. A team from McKinsey noted the phenomenon:

> When Prodigy entered the on-line services business, its management assumed that its chief value would lie in giving consumers

access to various kinds of published content—news reports, sports
scores, reference material and so on—as and when required. As it
turned out, subscribers were much more interested in communicating with one another.[6]

In other words, as it becomes easier and less expensive for individuals
to communicate with other individuals electronically, people are choos-
ing to chat longer. Quite possibly, the pleasures of human contact will
eat into the time that people now spend watching packaged electronic
entertainment. Social life will thus be enriched, not impoverished.

Virtual communities

Chatting to strangers becomes easier when based on a shared interest.
An important impact of the new forms of communications will be to
create new social bonds—"virtual communities"[7] of people linked electronically who meet occasionally for what cyber-enthusiasts call "face
time"—or who may never meet. These communities will grow up
around work interests, domestic interests, and cultural and ethnic
interests. Communities of interest, groups of people perhaps scattered
around the world, may have more in common with one another than
with their next-door neighbors.

Work interests
Electronic communities of interest have long existed in academic life,
medicine, and science. Neurosurgeons, experts on Anglo-Saxon gram-
mar, and earthquake specialists may all be fairly rare beasts, working
on their own or with others in equally abstruse fields. But they usually
know their colleagues around the world from their publications, from
conferences, and now from the Internet.

Such horizontal contacts will become more important as companies
fragment. In addition, electronic contacts can be used to help plan campaigns that challenge specific organizations or their actions by building
cross-border alliances. Environmentalists in many countries, for example, may unite to fight a battle against what they perceive as a destruc-
tive proposal in one country that may be financed by a donor or bank in

another. Or trade unions in one country may seek cooperation from
their counterparts in another to take on a multinational employer.

Domestic interests

All sorts of other groups, such as parents of missing children and those suf-
fering from rare diseases, have learned to exploit the potential of global
communications. Bulletin boards have become increasingly specialized.
For instance, Parent Soup, set up by two mothers in 1995, aims to be an
electronic park bench. It offers a special niche for every sort of parenthood,
from adoptive parents to parents with only one child, and a live chat forum
called "Sanity Break." Such contacts encourage self-help groups to flourish. They make it easier for people to draw on the collective experience of
other individuals, rather than resorting to professionals for advice on matters from medication to child-rearing. The technology that so neatly
matches buyers and sellers in many different markets also offers opportunities for finding new friends with similar interests. Above all, these on-
line communities indicate that the most compelling content of the
electronic world is not the output of Hollywood but simply—other people.

Cultural and ethnic interests

Electronic cultural and ethnic communities will strengthen ties that
distance might otherwise have frayed. The Internet provides a meeting
place for people separated from their native lands. Scotland's clans,
those strange and anachronistic Highland tribes, have been given a
new lease on life by Mel Gibson in *Braveheart* and by the Internet.
MacLeanNet, for example, a group of fifty or so members of the Clan
MacLean, provides a community in which members to e-mail one
another from cities as far from the glens as Dallas, Durban, and Mel-
bourne. North American Scots can find electronic listings for local
Highland Games (displays of ancient sports such as putting the shot
and tossing the caber) or can buy on-line a kilt to wear to them. They
can even call up a description, with photographs, of the proper way to
wear a sporran and the right place to stick a kilt pin.[8]

One of the most basic building blocks of cultural identity, a knowledge of one's origins, has been transformed by electronic communications, and especially by the Internet. To research a family tree once

meant a letter or a visit to a registry office, perhaps in another country. Now, several on-line genealogy forums explain how to track down ancestors. The home page of Cyndi Howells, a housewife from Washington state, for example, offers more than 17,400 (free) links to sources of genealogical information.[9] Translation software allows people to correspond with distant relatives whose language they cannot speak.

Inexpensive international telephone calls combined with the Internet allow exiled peoples to continue to practice their native language. Homesick emigrants will increasingly be able to shop from a distance for delicacies from home, watch television programs originating from their home countries, and read their home newspapers and follow their home sports teams.

In these ways, electronic communications will reinforce cultures that might otherwise be damaged by distance.

Language and Culture

. . . . .

The stock-in-trade of electronic media will be language and ideas. With the death of distance, many countries fear the power of American culture and the English language. They worry that their own languages will be swamped, and their cultures and traditional industries overwhelmed. Both fears are largely unfounded.

Electronic media affect language in three main ways: they alter the way language is used; they create a need for a global language that will most likely be filled by English; and they influence the future of other languages. In addition, the changes taking place in electronic media will lower the entry barriers to cultural industries such as television and movie-making.

A new linguistic style

Electronic communications have been changing the use of language for more than a century. To the delicate social question of how to answer the telephone, Alexander Graham Bell came up with the following solution: simply say, "Hello." Developments in telecommunications have brought other cultural innovations. For instance, the telephone made something that had been a rarity—a conversation with an unseen person—into a common experience. The telephone answering machine or voice mail produced new versions of the monologue (and one, to judge by the messages most people leave, that most of us do not much enjoy delivering). Radio sports commentary created what one American linguist describes as "a monologue . . . directed at an unknown, unseen, heterogenous mass audience who voluntarily choose to listen, do not see the activity being reported, and provide no feedback to the speaker."[10] Electronic mail and Internet chat have produced yet another linguistic innovation: the written conversation. Unlike the letter, to which a reply takes at least forty-eight hours (although in Victorian England responses were much faster), an e-mail reply can be more or less instantaneous. And e-mail has encouraged a vast number of people, many of whom may hardly have written a personal letter in their lives, to correspond. The revival of writing, even if only electronically, is surely a cultural trend to be welcomed.

The coming global tongue

For the electronic media to work efficiently as global carriers of language, they need a common language standard. Standardizing makes communication easier and cheaper. Like lower telephone tariffs and Internet charges, it creates a virtuous circle: communication is less expensive, so more of it occurs. English has emerged as the necessary standard.

English will soon be spoken by more people as a second tongue than as a first. (See Figure 9-1.) Until now, its spread has depended on two factors: the legacy of colonialism and the emergence of the United States as the world's largest commercial power. Most of the new countries that have emerged in the past half century have given English a special role, making it the dominant or the official language in more than sixty countries.[11] Otto von Bismarck, Germany's famous chancellor, foresaw the American commercial hegemony a century ago. Asked by a journalist in 1898 what he thought was the decisive factor in modern history, he replied "The fact that the North Americans speak English."[12]

In the future, the spread of English will be driven by two additional factors: the United States is easily the world's largest exporter of intellectual property (Britain is the second largest), and English is overwhelmingly the

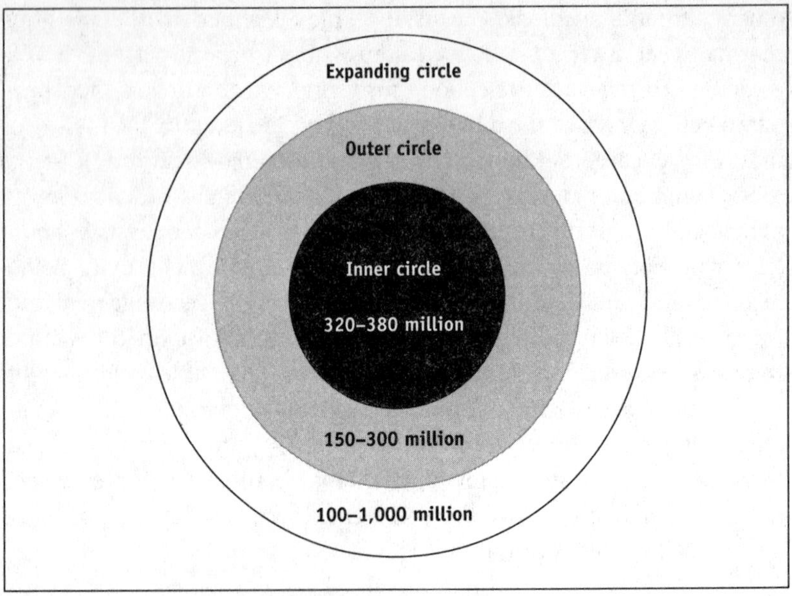

Figure 9-1 The Three Circles of English Speakers

The spread of English is shown here, visualized as three concentric circles, each representing different ways in which the language has been acquired and is used.

- The *inner circle*: English as the primary language. This includes the United States, United Kingdom, Ireland, Canada, Australia, and New Zealand.
- The *outer circle*: English as a part of a country's chief institutions, and an important second language. This includes Singapore, India, Malawi, and more than fifty other territories.
- The *expanding circle*: English taught as a foreign language and seen as an important international language, but not given any special status in language policy. This includes China, Japan, Israel, Greece, Poland, and a steadily increasing number of other countries.

Note: Figures refer to populations of English speakers.

Source: Adapted from David Crystal, *Cambridge Encyclopedia of the English Language*, (Cambridge: Cambridge University Press, 1995).

main language of the Internet. Computer software provides an example of the first: by some estimates, 80 percent of the information stored in the world's computers is in English.[13] People who buy intellectual property, whether Madonna or Microsoft, often buy English as part of the package. This is especially true of the Internet. One study, in 1996, found that almost all the scientific material on the Internet was in English; overall, the proportion of English to other languages is around 70 to 80 percent.[14]

The dominance of English on the Internet follows inevitably from the dominance of Americans among Internet hosts and users. In the future, as the proportion of non-native English speakers using the Internet rises, other languages will be more widely used. But English will most likely remain disproportionately important on the Internet, creating a large new category of English users: those who can write the language colloquially but cannot necessarily speak it.

The only two alternatives to the spread of English will be the rise of an alternative language (conceivably, Chinese might one day be a candidate, but the lead of English seems impregnable for the moment) or advances in machine translation. For the moment, machine translation leaves much to be desired: see, for example, Figure 9-2.

As world trade in on-line services grows, the prominent role of English will give those who speak English an advantage over those who do not. India and Jamaica have built their data-processing industries partly on the use of English. In the Philippines, where workers have familiarity with American English and a high level of literacy, the country offers a deregulated telephone service and seeks to position itself as an Asian media center. Biggest of all will be the advantage to those of us lucky enough to grow up speaking English: providing, of course, that future non-native speakers of English can understand us.

Other languages

If English will be established as the world standard language, what are the implications for other tongues? Paradoxically, they may benefit, too.

Digital television, with its multitude of channels, makes it far easier and less expensive to make and distribute niche programming in minority languages. S4C, for example, Britain's Welsh-language channel, has helped to make speaking Welsh fashionable among the Welsh young and has promoted the emergence of Welsh rock bands, film-makers, and even cartoonists. The channel plans to rebroadcast by satellite to the many Welsh speakers in continental Europe.[15]

The Internet may also protect subsidiary languages, for two reasons. The limitless capacity of cyberspace means that languages do not compete with each other head-on, as they have done in the past on analogue radio and television. A Danish rock festival can advertise on the

Hans Schultz
Gross Dingsbums Firma GmbH

Dear Mr. Schultz:

What a stroke of luck it was to run into you at the industry conference in Strasbourg
the other day. Who would have guessed that just putting our heads together for five minutes over a cup of coffee would lead to such a brainstorm. I think you could safely say
that our backgrounds weren't exactly cut from the same cloth. But lock us in a room
together for 15 minutes, and things definitely begin to click.

I was hoping you could lend me a hand with another matter. While the standards we
discussed are conventional according to the needs of most markets, the technology in this
field is racing ahead like a charging bull. Is there any assurance that if we form a joint
venture, connectibility and, more important, full integration of our products will still be
achievable in a reasonable length of time?

Best regards,

John Smith
The Great Widget Group PLC

Hans Schultz
Gross Dingsbums Firma GmbH

Sehr Geehrter Herr Schultz:

Was ein Schlag des Glücks, den es in Sie an der Industrietagung führen sollte, in
Straßburg der andere Tag. Wer hätte geraten dies nur unsere Köpfe plazieren Zusammen
würden Minuten über einer Tasse Kaffee für fünf zu solch einem Geistesblitz führen. Ich
denke, daß Sie sicher sagen konnten, daß unsere Hintergründe nicht genau von geschnit-
ten wurden, d Gleiches Tuch. Aber verriegeln Sie uns in einem Zimmer zusammen 15
Minuten, und Dinge beginnen bestimmt zu klicken.

Ich Hoffte, Daß Sie Mir Eine gibt Mit Einer Anderen Angelegenheit Leihen Konnten.
While die Standards Wir erörterten sind konventionell entsprechend dem Bedarf der meisten
Märkte, d Technologie in diesem Feld macht ein Wettrennen voraus wie eine rennende
Bulle. Gibt es irgendwelch? Versicherung das wenn wir formen einen Joint Venture,
Connectibility und, wichtiger, Volle Integration unserer Produkte, wird immer noch in einer
vernünftigen Länge der Zeit erzielbar sein?

Beste Grüße,

John Schmied
The große Widget Gruppenag

Hans Schultz
great what'sit bang company company with limited liability

Dear Mr Schultz:

What a blow of the luck, this one it should lead the other day out of you at the industry conference over street castle. Who would have looked after, that, put our eads about a
cup of coffee only five minutes, supply a spirit lightning suchly would. I think that you
could for certain say that our backgrounds exactly weren't cut by the same cloth. But lock
you for us in a room with 15 minutes and things starts clicking certainly.

I hoped that you could borrow a hand with another matter. We discussed the standards lasting conventional according to the need of most markets is, the technology makes
in this field a race like one reindeer end bull ahead. This is very insurance this if we form
one joint ventures, Connectibilty, and full integration of our products still will more importantly be obtainable in a sensible length of the time?

Best greetings,

John Schmied
den die gro

Figure 9-2   The Inadequacies of Machine Translation

This mock letter was translated by a software program that was instructed to trans-
late its own work back into the original language.

Source: *Convergence Magazine, Wall Street Journal Europe.*

World Wide Web in English, German, and Swedish—but it can also market itself in Danish. The Internet also offers an inexpensive way for speakers of exotic languages to keep in touch with each other. One description of the World Wide Web found "discussion groups in more than sixty languages, at which point I stopped counting." This explorer discovered people conversing in Aragonese, Armenian, Basque, Breton, Cambodian, Catalan, Esperanto, Estonian, Gaelic, Galician, Hindi, Hmong, Macedonian, Swahili, Welsh, Yoruba—and so on and so on.[16]

Cultural protection

Some countries, especially where English is not the first language, dread the erosion of their own cultures in the face of an onslaught of American products. They particularly fear that their own film and television industries will be overwhelmed by the output of Hollywood.

The fear is greatest in continental Europe. Two trends worry Europeans: the collapse of the local movie industries and the rise of American imports. One study of the box-office take of European-made films in their home market suggests that, by the early 1990s, it had dropped to one-sixth of its 1957 level in real terms. Between 1980 and the mid-1990s, the audience in Western Europe for locally made movies collapsed, falling from 475 to 120 million. Meanwhile, the audience for American-made movies barely changed. American films take 80 percent of box-office receipts in much of Western Europe, and the countries of the European Union take 60 percent of American film exports.[17]

Making a Hollywood blockbuster is extraordinarily expensive: a typical Hollywood movie costs $40 million to produce and another $20 million to market and advertise. Hollywood can afford its vast costs because it has built a vast global reach. Moreover, just as the City of London can dominate shipbroking or Silicon Valley the computer industry, so Hollywood has a critical mass of talented scriptwriters and large studios. Interestingly, all three businesses thrive on physical proximity: even though all three are particularly well suited to distance-working in theory, in practice, they all need the lunch culture.

By contrast, European studios make seventy or so feature movies each year with average production budgets of $4 to $5 million. (Another 230 or so European films are made for less than half that

. . . . .

Table 9-1   Average Results per Film in Europe, Japan, and the United States, 1993–1995, in Current U.S. $

	Cinema	Video	Pay TV	Free TV
Hollywood films				
Audience, m	10.5	34.0	24.0	147.0
Revenue, $m	58.0	103.0	17.0	22.0
European films				
Audience, m	0.1	0.2	0.3	1.5
Revenue, $m	0.8	0.7	0.2	0.2

Source: Reproduced from Martin Dale, *The Movie Game: The Film Business in Britain, Europe and America*, 1997 by permission of Cassell, 125 Strand, London, England.

amount.) But approximately 80 percent of the production costs of a European film are met by state subsidies, and this, not Hollywood clout, has destroyed Europe's movie industry. Subsidizing filmmaking encourages filmmakers to ignore audience appeal. The most extraordinary example appears in Germany, where half the revenues of the average movie come from the public purse, and almost half the films made between 1985 and 1991 were never shown to a paying audience. But the problem is continent-wide: "The majority of European films have virtually zero box-office appeal," says Martin Dale, author of a recent study of the industry.[18] (See Table 9-1.)

Some European politicians have tried throughout much of the 1990s to get the European Union to impose trade controls on American entertainment products. During the World Trade Organization negotiations on telecommunications between 1996 and 1997, the EU tried to exclude teleconferencing from the trade accord, for fear that it would be used to import video entertainment. So it may well be; but there is a better answer than prohibition.

To start with, Europe must accept that it will never develop a rival Hollywood (although Manila or Hong Kong may do so, and Bombay's "Bollywood" already produces more movies than any other city on earth). Blockbusters will remain a predominantly American phenomenon. But the effect of the electronic revolution will be to increase enormously the output of locally made entertainment, for two reasons. First, television viewers, unlike movie audiences, generally seem to prefer watching locally made fare. Europe will now have hundreds—even thousands—of television channels to fill. Already in the early 1990s,

European countries were becoming more willing to buy each other's television series.[19] The trend may accelerate: as the battle for audiences increases, the price of high-quality American-made movies on television will rise, and viewers may end up paying a premium to watch Hollywood's output. Second, digital equipment dramatically reduces the cost of making movies and television programs. The new digital-video camcorder format can film movies good enough to broadcast. Prices are falling dramatically for editing and other post-production equipment, as well. True, the big costs of Hollywood movie-making include multi-million-dollar payments to famous stars, directors, and scriptwriters as well as the immense expense of global marketing and distribution. But lower production costs will make it easier to foster local talent.

A likely consequence of these developments will be the growth of a second-tier industry of national movies and television programs. This industry will not produce many films like *Trainspotting*, let alone like *Four Weddings and a Funeral*. Few will be good enough to export to a world market, although some will enjoy a regional success. But digital broadcasting will allow people to choose the language of the soundtrack for whatever they watch, and digital manipulation of filmed images will improve the quality of dubbing. If France and Italy were once again to become centers of cinematic excellence, producing the likes of *Les Quatre Cent Coups* and *La Dolce Vita*, the pressure for protection would probably fade away.

Winners and Losers

. . . . .

The new world of electronic communications will include winners and losers, haves and have-nots. Governments have already begun to fret about the danger that some groups will be excluded because they are too poor to afford the equipment and gadgets of the telecommunications revolution. In rich countries, politicians worry about the possible exclusion of citizens with low incomes; in poor countries, politicians worry that the nation may not be connected at all. In May 1996, government ministers from forty-two countries, including thirty-two developing countries, assembled in South Africa to discuss the gap that they feared might open between rich countries and poor ones. In fact, this may be a revolution of inclusion and opportunity for many of today's have-nots.

In the rich world

In rich countries, becoming an electronic citizen will certainly require an entry pass: a link with the world's network of networks. But many of the worries about exclusion are wide of the mark, for two reasons. First, prices are falling at a staggering rate for almost all the machinery that connects people. Mobile telephones, set-top boxes for digital television, home computers, videophones: all are becoming cheaper not just relative to other products, but in absolute terms. In that sense, the poor have possibilities for electronic communications that did not exist five or ten years ago. Second, the cost of connecting people is tumbling. Telephone lines, wireless links, communications satellites, and television channels are also falling in price at an astonishing rate. Again, the result will be to bring to the poor facilities that might once have been a luxury. The only area where the poor may be excluded, for a while, will be in the upgrading of telephone and cable networks to carry high-speed computer links. Companies will concentrate first on well-to-do areas, where they expect to find more customers willing to pay for Internet access. But access will still be available to the poor, as to many other people, over a standard telephone line.

For some of the most excluded people in society, this will be a revolution of inclusion. Among those benefiting will be the housebound, both the old and the disabled. For them, anything that makes communications easier and less expensive will be a boon. The proportion of old people will grow dramatically in the rich countries: in some places, by 2010, one person in five will be over sixty-five. The lives of these people, many of whom will live well into their eighties, will be impoverished by the extent to which modern societies depend on the automobile. Once they become too old to drive, they will be far more isolated than they would have been in the first half of this century. For these people, many of whose relatives live too far away to visit frequently, inexpensive communications will be a lifeline.

Another group of beneficiaries of new communications will be the poor, who may find themselves more included in the new electronic society than they were in the old one. The U.S.-government sponsored program "Making Healthy Music" provides an example. Participants in low-income homes receive a computer, software, and training in exchange for their commitment to try to improve communications with their neighbors. They are then hooked into an intranet linking them to other local users. Some proponents of the program claim that it has helped participants to build friendships in neighborhoods where people fear venturing into the streets.[20]

In general, though, the uneducated and the old will benefit less than the well educated and the young. Those without education will find it hardest to make use of the many options becoming available. Access to all the books in the Library of Congress is of little use if you cannot read. Older people in general have to be "retrofitted," in a phrase used by the young consultants of Forrester Research in Cambridge, Massachusetts, to make the most of the new technologies that the young will simply grow up with. For at least a decade to come, many of the young will have a key advantage over their elders: a natural feel for the technology that is essential to the way offices will work.

In some countries, this alone will be a powerful force for social change: in Japanese companies, where age is tantamount to seniority, older workers suddenly find themselves forced to ask their juniors how to use the most important piece of office equipment, the computer; and junior workers find that they no longer need to pass a memo to a senior manager through their immediate boss, with all the opportunities for obstruction that provides, if they send their comments electronically. These new procedures will reinforce the trends that are already breaking down Japan's hierarchical employment patterns.

The young will grow up with the idea of almost limitless choice of entertainment, of easy access to information, and of the screen and the computer as gateways to the rest of the world. The electronic world will be dominated by the young.

In the developing world

As big an issue as the adaptation of rich countries to the electronic world is how to prevent developing countries from being excluded from that world altogether. Millions of people in developing countries have never

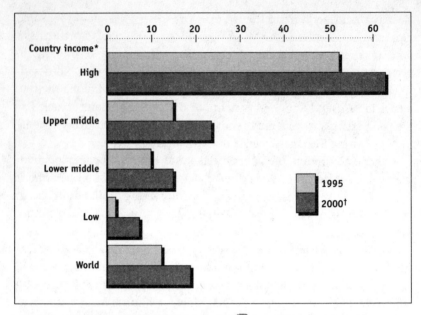

Figure 9-3   Telephone Lines per Population of 100 People, 1995–2000.
*World Bank income groups based on GNP per head: high $8,626 and over, upper middle $2,786–8,625, lower middle $696–$2,785, and low $695 or less.
†Based on growth rates over the previous six years.
Source: International Telecommunication Union.

made a telephone call, let alone used the Internet. (See Figure 9-3.) Indeed, in some African countries (Sierra Leone, Uganda, and Zimbabwe), the number of people has been growing faster than the number of telephone lines.[21]

A country's level of wealth has, up until now, generally been correlated with the number of telephone lines per one hundred inhabitants; wealth seems to be an essential condition for increasing communications facilities. Yet the effects of poverty are frequently compounded by badly designed telecommunications regulations and by restrictions on would-be private investors, both of which are often designed to protect an inefficient national telecommunications monopoly. In addition, investment costs per telephone line are often high.

These problems can be solved with appropriate policies. Once they are, developing countries face a great opportunity. As with the rich world, so with poor countries: the cost of access declines rapidly with technological advances in communications. Developing countries, therefore, have the potential to skip several stages of technological development and go straight to the most up-to-date network. Nearly all of the countries with largely digital telephone networks are in the developing world. In Japan, the most digitalized of the rich countries, only 72 percent of subscriber lines connected to digital exchanges in 1994, compared with 83 percent in Mexico and 100 percent in Chile.[22]

Connected to the world's networks, poor countries will have one enormous advantage. They are home to most of the world's young. If these countries allow their communications and media industries to flourish—if they liberalize markets, regulate the inevitable concentrations of market power, protect freedom of speech, and promote education and literacy—they will eventually be the biggest beneficiaries of the telecommunications revolution. A country such as India, with enormous creativity and widespread use of English, or Chile, with its relatively open telecommunications market, or China, with its extraordinary attention to education, might leapfrog ahead of many rich-world competitors.[23] Distance will no longer be an obstacle.

The most important effect of the death of distance will be to narrow gaps, not to widen them. Where countries adapt their policies to allow new communications industries to flourish, they will find that the electronic world creates opportunities rather than suppressing them and opens doors rather than shutting them. Communications, after all, is all about narrowing gaps.

10
Government and the Nation State

Half a century ago, in 1948, George Orwell described a world dictator-
ship built on electronic communications that could monitor its citizens'
every action. Now, when electronic networks can connect databases and
video cameras around the world to unprecedented computer power,
Orwell's vision in *1984* has become a much more practical proposition
than it was in his day. Big Brother really could watch you and know an
enormous amount about your banking, travel, and spending activities;
your television viewing and magazine reading tastes; your health worries; and the friends whom you telephoned last week.

Yet Orwell's harsh vision will not come about. The nature of the state
will undoubtedly change as a result of the death of distance, but not in
the way that Orwell expected. Certainly governments will acquire the
technological capability to isolate and track every movement of those
citizens it regards with suspicion, whether because of their ethnic origins, past behavior, political views, or religious or other beliefs. This
surveillance power will be a serious potential danger to liberty, especially as countries without Western respect for human rights acquire
Western networks and computers.

But a far likelier development will be a reduction in the authority of
the nation state. The death of distance will shift power downward, to
the individual. It will both reinforce democracy and transform it.

Historically, the world's nastiest governments have tended to have
bad communications and a majority of citizens too poor to leave data
trails. Is that an accident? Or is there, in fact, a strong link between
good communications and political freedom? Certainly dictatorial governments want to control what appears on television, as China has
done, and they often fear the Internet: in Myanmar, one of the world's

. . . . .

most unpleasant governments threatens any unauthorized owner of a networked computer with fifteen years in jail.[1]

Everywhere, however, free communications will change the balance of power between governments and their citizens. People will be able to become better informed—even though most governments are no more enthusiastic about putting information into cyberspace than they were about publishing it in more traditional ways. People will be able to communicate their views on their rulers more easily. This ability matters particularly for people living under unpleasant regimes, who have thus gained a new way to make their voices heard. And it will become possible to give people more power to participate directly in government decision, if that is what voters want. In the United States, these forces are shifting politics toward increased sensitivity to lobbying and to public-opinion polls. At a later stage, they will have potent effect in the many countries just beginning to experiment with democracy.

In the established democracies, where many people find their governments unresponsive and extravagant, the telecommunications revolution will affect the size and role of government in two ways. First, a stronger link will be forged between what voters want their government to do and what they are prepared to pay in taxation. As commercial activity becomes more mobile and less a prisoner of geography, some taxes will become almost voluntary. Second, the consequent reduction in tax yields will force governments to reconsider the ways in which they provide services. They will target services more precisely, cut costs, and concentrate on regulating and monitoring. Government will therefore become smaller. The telecommunications revolution will, further, affect the size of the nation state. Historically, the size of political units has mostly been determined by geography. In Napoleonic France, as Eli Noam of Columbia University has pointed out, the sizes of the *departments*, the main administrative unit, were established so as to permit officials to travel to any part of them and still be home in time for dinner.[2] The death of distance will not only erode national borders; it will reduce the handicaps that isolation has previously imposed on countries on the fringes of economic regions. The effective dissolution of geographical disadvantages will be enormously important not just for the many small countries that have come into existence in the past half century, but for independence movements around the world: the economic case against secession will be weakened.

Finally, the death of distance will be a powerful force for peace. The best way to discourage countries from fighting one another will be through communication. Not only will government leaders be better able to communicate, but ordinary citizens everywhere will also become more familiar with the ideas and aspirations of people the world around. The citizens of one country may come to understand those in other countries a little better, and the glue that binds humankind will be strengthened.

The Political Process

· · · · ·

Al Gore, the U.S. vice president, has talked of using communications technology for "forging a new Athenian age of democracy." That might not be such a good idea: after all, Athenian democracy excluded women and slaves, a majority of the population, from the rights of citizenship, and the Athenian assembly was notoriously prone to being hijacked by oligarchs and demagogues. But Gore's essential point remains valid: the changing technology of communications will clearly alter the relationship between politicians and the governed—almost certainly for the better.

That is not surprising. In this century, communications have probably been the biggest single force for transforming the way democracies work: allowing people with ideas to rally voters; allowing voters to be heard by politicians; and allowing politicians to argue with both camps. Radio and television have revolutionized both the way political debate takes place and the things politicians have to do to get elected and re-elected.

Now, free communications will change the balance of power between government and its citizens in several new, important ways. Citizens can, if they choose, become better informed; they can more easily make their views known; and they will be able, in theory, to vote on individual policy decisions, in a sort of continuous referendum.

Informing the citizen

Good information is essential for effective political involvement, and the communications revolution makes information more readily accessible than ever before. It also allows people a wider choice of sources of information, especially important in repressive countries where the national news media are biased or controlled by government.

Information that was once expensive to obtain can now be procured inexpensively. The Clinton administration, for example, insists that all federal public documents (including slide presentations made by officials) must be accessible on the Internet. Access to publicly available information is no longer confined to an elite (the media, officials, big business). It has become much easier for citizens to discover facts that affect their lives. For instance, a local group of environmentalists can now consult the Toxic Release Inventory (which lists some of the nasty emissions that chemical plants give off), search through it by zip code, and pick out the relevant figures for the local plant.

But use of communications technology (hacking excepted) will not help citizens root out facts that a government is determined to keep from them; and, although many more people may be able to read it, information from, say, the European Commission will not necessarily be more useful—or more comprehensible—simply because it is electronically available.

Gradually, better access to information may change the culture of bureaucracies. Politicians may become less frightened of making information widely and speedily available once they realize that the political consequences of doing so often entail less risk and damage than a policy of obfuscation.

Above all, people will have access to a diversity of information, which will help to keep governments honest. Government propaganda becomes less convincing once people hear the other sides in an argument. Until now, the main source of alternative news in many countries has been radio—mainly the BBC overseas service and, to a lesser extent, Voice of America, stations that survive using government grants. Now, the Internet creates a much less expensive source of alternative news, and inexpensive access to global sources of information will be a strong antidote to prejudice, nationalism, and war-mongering.

A voice for citizens

Whether they want to or not, governments will become more aware of what their citizens think of their policies. People living under repressive regimes now have new ways to make their voices heard; and in democracies, the Internet and other communications improvements reduce lobbying costs.

Better communications enable people living under repressive regimes to carry their campaigns for freedom to the outside world. When, in December 1996, the government of Serbia tried to shut down B-92, the last independent radio station, the station broadcast in Serbian and English on the Internet instead. Partly because the opposition to the government centered in the universities, where many professors and students had access to the Internet, the radio station's Web site played a crucial role in distributing news of the anti-government protests.[3]

That particular campaign ultimately succeeded. Not all do. For all the publicity that the Tiananmen Square rebels achieved by using the fax to communicate with the rest of the world, their rebellion was still quashed. Even with foreign television cameras rolling, the government was willing to use tanks and troops.

Lobbying politicians is universal in democracies, but electronic mail makes it much faster for enraged voters to send their views to politicians on impulse. A letter or fax, by comparison, takes time and preparation: at the click of a mouse, a lobbyist can send a message to every member of Congress. President Clinton, who receives one to two thousand items of e-mail every day, qualifies by far as the world's most e-mailed person. (Of course, when one gets that much e-mail, it is difficult to read everything. In 1994, Carl Bildt, then the Swedish prime minister and an inveterate nethead, sent an e-mail to President Clinton. It was a historic moment: the first Internet exchange between two such senior politicians. What happened? Nothing. After two days of waiting, Mr. Bildt's staff rang the White House instead.)[4]

Desirable though it is that individuals can communicate more readily with politicians, some people fear that the productivity of lobbyists has also been increased. A "broadcast fax" can now be sent to hundreds of sympathizers with one push of the button, and the mobile telephone allows the lobbyist sitting in a Congressional hearing to communicate with a lobbying group's head office. Jonathan Rauch, a columnist for the *New Republic*, argues that the liberation of lobbyists harms democracy by giving too much power to special interest groups.[5] Politicians, as a result, according to Mr. Rauch, become hopelessly in thrall to what they interpret as public opinion. Influenced by the hubbub of electronically delivered special pleading, they fail to think about the interests of the country as a whole.

However, if many more people clamor for the ear of politicians, each will drown out the others. Electronic arrival of a message may, in fact, devalue it. Some American politicians complain that 80 percent of their e-mail comes from people who do not live in the state or district they represent. The old-fashioned letter on paper, typed—or, better, laboriously hand-written—has, ironically, become more influential than the urgent screed on the screen.

Direct democracy

In the past two decades, democracy has become the standard form of government around the world. During that time, at least thirty-two formerly authoritarian regimes have held relatively free elections.[6] But voters often feel dissatisfied. Turnouts flag; voters defect from political parties to lobbying groups; extremists flourish. Such revolts against the democratic process, say some analysts, reflect a feeling among voters that their voices go unheard. The average citizen votes only every few years, usually with an infinitesimal influence on the outcome.

Now, some observers imagine that the Internet will encourage further transformation of political debate. In the United States, in particular, enthusiasts such as Alvin and Heidi Toffler keenly extol what they call "semi-direct democracy." The Tofflers argue in their book *Creating a New Civilization*[7] that voters should be allowed to make many more policy decisions. After all, if people can shop from home, no obvious technical reason prevents them from eventually voting from home.

A number of experiments using the telephone and television have been conducted over the past two decades, mainly in the United States, and mainly by Theodore Becker and Christa Daryl Slaton, political scientists at Auburn University in Alabama. They have developed the idea of "televoting" and of the "electronic town meeting," modeled on the New England custom of bringing a town's citizens together once a year to decide on the budget and other matters of moment. In televoting, television, radio, and newspapers set out the issues in a debate as fully as possible. A random sample of voters receives a pack of information, and these people are later contacted by telephone and, as Becker puts it, "facilitated towards a teleconsensus."[8]

Should televoting become armchair democracy? The supporters of direct democracy point to various positive consequences of such a development. Foremost is the possibility of higher election turnouts. In 1996, 92.8 million Americans voted in the presidential election; 94 million watched the Super Bowl. If it were as convenient to vote as to watch the Super Bowl, more people might do so. In addition, some form of televoting might help define constituencies more precisely. Already in the United States, the drawing of political boundaries attempts to ensure an appropriate racial mix; in the future, "virtual" constituencies could be created, drawn from voters randomly selected from all over a state or a country, or constituencies could be designed to represent different ethnic groups. It might be possible to eliminate elected representatives entirely: home after a day's work, people might be offered a slate of propositions on which to vote. Proponents of such a view imagine a government by the entire citizenry, electronically assembled. Such a system would reduce the power of lobbyists: it may be easy to persuade a few hundred representatives to a particular viewpoint, but it would be far harder to persuade an entire electorate.

Can direct democracy live up to the claims made for it? The possibility interests social and political theorists in particular because of the recent rise in referendums, a traditional form of direct democracy. Since the start of the 1970s, the frequency of national referendums has almost doubled.[9] Many have been held in the United States, where almost half of all states allow them. American referendums have an unusual characteristic: they are held mainly in response to voters' propositions or initiatives. A lobbying group with the requisite number of supporters can, in many states, put a proposition to the electorate at large. The Internet makes it easier to muster support for a proposition.

But mustering support is one thing; allowing voters to take binding decisions with the click of a mouse or the push of a button on a television remote control is quite another. Several problems would have to be resolved. First, the wording of any referendum must be clear and unbiased. Opinion pollsters know that the answer to any given question depends on the wording. Over a one-year period in the mid-1990s, British support for membership of the European Union fluctuated between 10 percent and 60 percent of those polled, partly because of small changes in the wording of the question.

Second, voters would have to decide whether they trusted their fellow citizens to make sensible decisions or thought they would be too impatient or shortsighted to take the trouble to understand the complexities of certain issues. Many of the decisions made by elected representatives involve complex and sometimes fuzzy arguments, rather than the simple yes-or-no positions required by an opinion poll or referendum. Government runs on compromises and horse trades, a process which irritates many voters and provides lush grazing for lobbyists, but which ultimately fosters consensus and moderation.

Third, society would have to decide whether a political system, operating on the assumption that the majority view should prevail, can also incorporate checks and balances adequate to protect the rights of minorities. That question would loom larger if those most in need of government protection, the poor and unschooled, turned out to be less likely to vote from their armchairs than would the better off. Studies in Switzerland and the United States suggest that turnout for referendums always falls below that for candidate-choosing elections and that the proportion of the vote cast by the poor and uneducated declines as turnout falls.[10] (See Table 10-1.)

Fourth, voters must really want teledemocracy. One relevant study looked at a series of ballots on various propositions held in New York in 1976 and 1977; these were publicized on television, and viewers were encouraged to clip ballot papers from newspapers and use them to cast their votes on the issues. Results showed that, initially, around 10 percent of the audience did so, but that as the novelty faded, participation declined to a mere 1 percent.[11] The viewers in that case knew that their views were not constitutionally binding. Televoting enthusiasts argue that popular enthusiasm would soar—and would be sustainable—if the mechanism had a formal role in the political system. But the evidence of low turnout in referendums around the world suggests that may not be the case. Even in Switzerland, which has held half the national referendums conducted in this century, turnouts since the 1960s have averaged no more than 45 percent. Despite its irritation with politicians, the electorate clearly is not keen to take over their job.

Could it be that many voters prize the representative aspect of democracy? Voters may prefer to delegate most of the business of running the country to specialists, as they delegate the business of, say, educating their children or finding a selection of holiday destinations. Just as electronic communications may alter, but will not destroy, the

Table 10-1   Average Percentage of Electoral Turnout, 1945–1993

Country	Candidate elections	Referendums	Difference
New Zealand	90	60	-30
Austria	93	64	-29
Sweden	85	67	-18
Italy	90	74	-16
Switzerland	61	45	-16
Ireland	73	58	-15
Britian	77	65	-12
Denmark	86	74	-12
Australia*	95	90	-5
France	77	72	-5
Norway	81	78	-3
Belgium*	92	92	0

*Compulsory voting laws in place.
Source: David Butler and Austin Ranney (eds.), *Referendums Around the World*. Reprinted with the permission of The American Enterprise Institute for Public Policy Research, Washington, D. C.

role of intermediaries in commerce, so they will not kill off representative democracy. The political intermediary—the politician—will remain, performing a specialist role on behalf of voters reluctant to carry the burden of deciding everything from the size of the state budget to the appropriate weight limits for trucks.

But even if voters do not want to make political decisions directly, representation will increasingly become a convenience, rather than a necessity. Thanks to advances in communications and information technology, the business of informing voters and seeking their views will become more sophisticated. So will the business of lobbying them. And the more political power shifts from politicians to the people, the more interest groups will have to lobby the people, rather than their representatives.

The Shrinking of the State

. . . . .

As the communications revolution alters the relationship between voters and governments, it facilitates a new role for the state: one of informing, monitoring, and measuring services rather than providing them directly. Just as companies will become smaller and more fluid organizations, employing fewer people directly, so government will increasingly become a coordinator rather than a service provider. The main pressure for change on governments will come from the erosion of tax revenues. The erosion will be greatest for indirect taxes; but direct taxes and corporate taxes will probably be affected as well.

In response, governments will hunt for ways to reduce state spending. Electronic communications and information technology will make it easier to contract out some services; target other services more precisely; and raise the productivity of those services that the state continues to pay for.

The state will employ fewer people and spend less revenue, shifting its role from provider to promoter and regulator. But if the primary task of the state becomes monitoring the delivery of services, more information than ever before will be required about citizens' behavior. Governments will need to know a great deal about entitlement and about the quality of delivery. One price of smaller government may thus be a loss of privacy.

Taxation

When businesses can sell goods and services across borders by mail order or on-line, which country taxes them, that of the supplier or of the customer? More to the point, how is the revenue to be collected? Tax officials in all industrial countries are worrying about these issues. The U.S. Treasury Department, for example, produced a consultative paper in November 1996 on some of the implications for tax policy of electronic commerce, suggesting that a study be conducted as to whether foreign businesses, physically located outside the United States but trading with American customers, should—or could—be taxed as being engaged in an American business. The OECD, a club of the twenty-nine richest industrial countries, is developing guidelines to help member countries police taxation on the Internet. The European Commission is doing much the same.

Threats to tax revenues come from both illegal tax evasion and legal tax avoidance. One effect of cross-border commerce will be to encourage evasion of value-added tax (VAT) and sales taxes. In general, taxation is more easily levied on physical products sent to a customer than on services. That is particularly true when goods are purchased from foreign providers. Consulting a database held on a foreign computer or a doctor located in another country, or buying electronically delivered software may be extremely hard to monitor, let alone tax. Tax evasion will be exacerbated if electronic money becomes widely used and internationally acceptable.

The death of distance will encourage legal tax avoidance in at least two ways: the ability to set up companies anywhere in the world will make it harder to levy corporate taxes, and the freedom for some people to work from anywhere in the world will make it harder to collect income tax, especially on high earners.

Illegal tax evasion will become harder to tackle, in at least four ways. First, the identity of someone making an Internet transaction is easy to hide—so people could, for instance, pretend to be operating from a foreign country when really they were not, thus evading local taxes. Second, electronic money transmission increases the ease with which an account can be opened abroad and money transferred into it anonymously, without the depositor ever traveling to the bank itself. One Web site allows a user to open a bank account in Antigua; the bank, which promises secrecy, offers numbered accounts, international wire transfers, and tax protection. Third, the creation of electronic money will eventually allow new opportunities for tax evasion. Some proposed schemes will, like cash, leave no data trail. But while with transactions in real cash, tax investigators can check purchases of stocks against recorded sales of output, on-line commerce will make such cross-checking much more difficult. If a taxpayer sells computer software on floppy disks, for example, the business's purchases of blank disks can be checked against recorded sales, and a tax inspector will sight obvious discrepancies. The same software sold electronically will produce no such record.[12]

Fourth, and most important of all, electronic commerce will encourage lots of people to try to evade indirect taxes. The biggest single source of indirect tax revenue in all industrial countries, except the United States and Australia (which both have sales taxes), is the value-added tax. VAT, levied at rates ranging from 3 percent in Japan to 25 percent in Sweden and Denmark, raises roughly 18 percent of all tax revenue in the OECD.[13] Most OECD countries will have the same prob-

. . . . .

lem collecting VAT as the American states face in trying to collect sales taxes for on-line transactions. But VAT is more complicated, because it is supposed to be collected at each stage in the production process, and then rebated, in order to levy the tax only on the value added at each point along the way. Rules exist for applying VAT to mail-order products, at least in the European Union, where the company that makes the sale is supposed to pay VAT to its own government—at the rate that would have been imposed in the country of the recipient. If on-line marketing leads to a surge in cross-border commerce, this complicated arrangement could easily break down.

Already, it has become clear how easily customs procedures can be overwhelmed. An ominous sign of things to come arose in the Netherlands in December 1996, when small foreign firms began offering CDs over the Internet, at prices well below those charged in Dutch music stores, and mailing them without customs declarations in order to avoid VAT. The only way to prevent this was heavy-handed and disruptive. The Dutch government therefore ordered the post office to open all packages that looked as though they might contain a CD. Dutch mail orders for CDs promptly collapsed.

Even countries that have clear rules on the subject tend to rely on the individual to pay voluntarily. In Australia, where music CDs at retail cost almost twice what they cost in the United States, the local sales tax is 26.4 percent. But since Australian Customs do not collect tax payments of less than A\$50 (about \$40) on personal imports, CD buyers can legally order about eight discs from abroad over the Internet without paying sales tax. Those who buy more than eight risk little, however, given the near impossibility of tracing the crime. In Canada, goods downloaded electronically attract VAT, but customers are on their honor to pay. Few are likely to do so. In the United States, because of a legal loophole, Americans rarely pay sales tax for mail-order or toll-free telephone purchases from companies based in other states, saving them nearly \$4 billion in state sales taxes annually.[14] The Supreme Court has ruled that residents who make electronic purchases from vendors in other states may be taxed by the state in which they live, but the vendor is not required to collect taxes. In addition, electronic commerce raises complicated questions about where a transaction occurs. Plenty of on-line shoppers therefore pay nothing. For some states, such as Texas, which gets more than half its total tax revenue from sales tax, the tax implications of electronic commerce are a serious threat.[15]

Legal tax avoidance may be almost equally intractable. A fundamental requirement for most taxation, especially on income and sales, is knowledge of location: the place where a person is resident, the national origin of the income a company receives, the place where a transaction takes place. Electronic commerce not only allows people and companies to sell services in one country that are bought in another; it also makes it harder to decide where a particular transaction has taken place.

Taxing companies has already become more difficult as it has become easier for some of them to choose where to pay tax on the various stages of their production process. Now a whole new group of industries and small businesses will be able to make the same choice. Communications technology will thus exacerbate the problem.

In addition, high earners will have more freedom to decide to live where they choose. Governments have already begun to cut top tax rates: every important industrial country had lower top tax rates in the mid-1990s than at the start of the 1980s. Professionals in countries where top tax rates remain relatively high, such as Germany, increasingly move to work in countries where tax rates are relatively low, as they were in Britain in early 1997. Businesses that employ large numbers of highly paid professionals will migrate to low-tax countries. As a result, countries that up to now have tended to compete for inward industrial investment will increasingly compete in a global market to attract inward professional and management talent. The winners will be those countries that can offer the combination of the lowest taxes, especially on incomes, and the highest quality of life.

To offset the coming losses in tax revenue, a group of European economists, set up by the European Commission, has proposed a "bit" tax, the brainchild of Luc Soete, professor of international economics at the University of Limburg in the Netherlands, to be paid by Internet Service Providers and levied on the bits—effectively, the amount of information—downloaded by individual Internet users. The Commission has no powers to impose a new tax on member countries, and most governments, worried about Europe's lack of competitiveness in information technology, are unlikely to adopt the scheme. Imposing a bit tax is a sure way to encourage those bits to walk.

. . . . .

U.S. President Bill Clinton, by contrast, has suggested a duty-free zone for all goods and services delivered electronically. That idea, guaranteed to endear him to cyber-surfers and to encourage the growth of Internet commerce, probably lies in the U.S. interest, since the United States will be, for some time to come, a big net exporter of such products.

In other ways, however, the combination of communications and computers may create opportunities for new sources of revenue. The electronic revolution makes it possible to charge for small increments of things, such as the use of each hundred yards of congested city-center roads or pollution from individual cars, and to vary that pricing in all sorts of ways, such as charging people more for electricity at periods of peak demand or for water used at a rate of more than a certain number of gallons per day. Such environmental pricing may well provide a rising share of government tax revenues in the future.

State spending

Faced with eroding tax revenues, governments will look for ways to reduce spending. Reductions in spending will come from outsourcing as many services as possible, improving the targeting of services that the state continues to provide, and increasing the efficiency of the services for which the state continues to pay.

Welfare payments

The combination of databases and smartcards (pieces of plastic that store information on a microchip) with mobile communications allows government bodies both to reduce fraud and to verify transactions carried out by the private sector on its behalf.

In South Africa, for example, the payment of state retirement pensions in remote parts of the bush is being transformed by the combination of satellite technology and touch-screen automated teller machines (ATMs). At present, most state pensions are paid by check. In villages where there are no banks, checks often must be cashed at a local shop, which means the pensioner pays a commission. Many pensioners, moreover, are illiterate, making it difficult for them to know if they are being cheated. Conversely, the provincial governments paying the pensions also lose money—one estimate suggests around $2 billion per year—because many people make fraudulent claims.

First National, South Africa's biggest bank, has devised a procedure, subsidized by the provincial governments, through which pensioners receive a bank card containing an ordinary magnetic strip along with an electronic record of their index-finger print. To receive their money, the pensioners insert their bank cards in a cash dispenser and put their index fingers against an electronic scanner; an ATM delivers the cash only if the print matches the card. For rural areas, the bank straps a mobile ATM to a truck that visits different villages on a fixed day each month—pursued by an entourage of migrant traders offering everything from chickens to cloth. The system has also been adopted in neighboring Namibia.[16]

In the United States, an electronic transfer system for paying welfare benefits has been installed in the state of Texas by Transactive, a subsidiary of GTECH, a Rhode Island company. Instead of food stamps, Texas's three million claimants receive a "Lone Star Card." When they enter a PIN number and swipe the card through a terminal at the check-out counter of fifteen thousand retail outlets, the cost of their food purchases is deducted automatically from their entitlement and added to the retailer's account. The system recognizes the bar-codes of forbidden items such as alcohol and cigarettes, thus ensuring that food stamps are actually spent on food. Texas officials noticed that, when towns introduced the new procedure, sales of alcohol tended to fall and sales of food to rise. By weeding out fraudulent claims, the system has produced dramatic savings. Claims have fallen by 10 percent, and savings are running at about $37 million per year.[17]

Health care

After welfare, health care is the largest single item on the budgets of most rich countries. Even in the United States, publicly financed health care absorbs a larger share of GDP than does defense. The burden will rise with the aging of the population, forcing governments to look for imaginative ways to reduce costs. The easiest single way to cut costs is to keep patients out of hospitals and to use the time of trained staff in productive ways.

- - - - -

That lesson has already been learned in Guyana, one of the pioneers of long-distance primary care. In a country where experts are few and mainly clustered in central hospitals, primary health care has to be given mainly by local workers with only a few months' training. Rural health workers in Guyana, having had about a year of special training, use a two-way radio network to check on the delivery of drugs and supplies and to ask advice on problems they are not equipped to solve.[18]

Now, nurses at the Hays Medical Center in Kansas are trying similar techniques. Using telephones and computers, they monitor elderly patients living in the vast spaces of the western Kansas plains. The system allows a nurse to check up to fifteen patients an hour, while a nurse physically traveling from house to house may see only five or six in a day. Each televisit costs $36, compared with $135 for a home visit by a nurse or $60 for a visit by a nursing assistant.[19]

Interest in telemedicine in the United States has grown rapidly in the mid-1990s. Given that America's nurses and health aides make around five hundred million home visits to patients each year—and that the number will rise in line with the number of elderly folk living at home—the potential savings are enormous. The Federal government is considering paying a lump sum for each Medicare patient's home care, rather than paying for each visit. Telemedicine, backed up by home visits, is now being used to monitor patients with diabetes, hemophilia, skin disorders, and a whole range of other ailments.

Another electronic health care technique involves devices with robotic voices that ask patients at regular intervals whether they have taken their medication. If the patient fails to touch the "yes" square on the screen, the machine eventually contacts a nurse, who telephones the patient to ensure the pills are taken. That will bring savings too: failure to take medicine at the right times is a leading cause of hospital admissions.[20]

Eventually it may be possible to save on hospital costs by carrying out operations at home, using the skills of doctors many miles away to carry out remote surgery. That would also allow small hospitals to save money by drawing on the skills of a distant pool of surgeons. The techniques are being developed by the Pentagon, its interest driven by the need to find better ways to operate on wounded soldiers in battlefields where doctors cannot safely go. In past wars, up to 90 percent of the deaths of those who were injured in frontline battle occurred because the wounded did not reach a hospital in time. By the end of the century, the American Army Medical Department hopes that it will be possible to bring a mobile operating room to the wounded. A surgeon, far away but watching on a three-dimensional monitor, would operate using a central master control that would manipulate remote forceps, scalpel, and needles. What may become possible for a wounded soldier will also, in time, be available for civilians.[21]

Education

The expense of higher education gives new impetus to distance learning, especially at the university level. Distance-learning is not new: Britain's Open University has been offering courses over radio and television networks for twenty-five years, while in the United States, more than thirty programs offer university courses to people in isolated regions, mainly in engineering. The oldest distance-learning university of all is the University of South Africa, at which both Nelson Mandela of South Africa and Zimbabwe's Robert Mugabe earned their degrees.

In poorer countries, teachers often account for 90 percent of the education budget, so the easiest way to cut costs is to increase class size. China has taken that to the extreme: the Central China Television University has one to two million students, more than the rest of the world's distance-learners put together. The presentation is said to be fairly utilitarian: a television camera is pointed at a lecturer, and the result is relayed by satellite to classrooms all over China. But for many countries, a similar program may be the only alternative to no higher education for many people.

Where communities are scattered, distance learning offers particularly big savings. The University of the South Pacific has a satellite-based network linking its main campus in Suva, Fiji, with its agricultural college in Western Samoa and other centers in nine Pacific-island nations. The result has been savings in staff travel time and costs and a decline in dropout rates for students.[22]

Even in rich countries, more and more higher education and training will probably be delivered long-distance, either to special classrooms in companies or at other locations, as is typical in the United States, or to students in their homes (and, in the case of some, prison cells), as in the British and South African tradition. Techniques being developed by companies such as Hewlett-Packard and Xerox will spread to campuses only as parents grow restive at the cost of university education. In 1995, private universities in the United States effectively charged each student nearly $60 per lecture—a figure that will have since increased and that did not take account of either public or private support. Yet one survey of consumers found that only about 31 percent felt that college tuition gave "good" or even "average" value for the money.[23] Distance learning may not have the cachet of a good university name, but it will be as good as, and less expensive than, a mediocre one.

Crime prevention

When sixty remote-controlled video cameras were installed in Kings Lynn, Norwich, crime fell almost immediately to one-seventieth of its previous level. The savings in patrol costs alone rapidly paid for the equipment. Today, more than 250,000 cameras are in place near trouble spots around the United Kingdom, transmitting information round the clock to one hundred constabularies, with a result for most of a fall in public misconduct.[24] A study of the effect of cameras in Newcastle found that arrest rates also rose after they were installed, by a quarter for burglary, criminal damage, and offenses of drunkenness.

A combination of security cameras and computers will also be a low-cost way to improve driving discipline. Installing cameras routinely at junctions will eventually allow police to issue tickets automatically to those who jump red lights or exceed the speed limit. The effects of using security cameras will only improve as it becomes less expensive to employ people on the other side of the world to watch them overnight, alerting a local guard whenever they see something suspect.

Form-filling

One of the main improvements in the productivity of the public sector will be a reduction in form-filling and an advance in the accuracy with which information is recorded. As part of a federal scheme to improve access to government services in the United States, the Social Security Administration has been experimenting with making benefit statements available electronically. The first results suggest that the SSA saved more than one dollar for each statement issued on-line—and individuals saved a wait of up to three weeks. Another scheme allows authorized SSA employees to look at state records on-line, which speeds benefit payments by up to a week.

Taxpayers in the United States can download tax forms directly from the Web site of the Internal Revenue Service: the site had almost three million "hits" on the day before forms were due in 1997, probably the largest ever recorded on a single day.[25] In time, it will be possible for individuals to file tax forms electronically (it can already be done by an adviser such as an accountant).

Under another scheme being devised in Britain, self-employed workers will be able to use the Internet to submit forms giving details of their social-security contributions. At present, self-employed workers submit 350,000 forms a year. For the quarter of that number which are incorrectly completed, government staff must make time-consuming telephone calls to the workers, return the forms for revision, and then manually insert the changes on the forms. In the course of 1997, Microsoft with Electronic Data Systems is developing a form that can be electronically submitted, and software that will pick up errors automatically.

Border control

In October 1996, IBM signed a deal with the Bermuda International Airport to pioneer its "FastGate" immigration smartcard, aimed at slashing passport queues. Passengers can apply to encode passport details and an identifying handprint on a credit card or perhaps a form of airline frequent-flyer card. On arriving at an airport, passengers will swipe the card through a machine and place their hands on an electronic reader. IBM thinks the system will become standard at airports everywhere within five years.[26]

Already, British passengers flying to Australia can be cleared electronically for entry before they leave their home country airport. Something similar is on trial for shipments of goods crossing from the United States into Canada and Mexico, its two partners in the North American Free Trade Area. Until now, truck drivers have had to present a sheaf of documents to customs officials at the border, causing long delays. In the trials, companies transmit information to the border posts over the Internet, and the trucks use wireless links to tell officials when they are about to arrive.[27]

. . . . .

Exporting services

Once public services can be provided remotely, they can also be exported, generating revenue—or imported, cutting costs. Under a program at the Massachusetts General Hospital, for example, a team of seventy radiologists has x-rays wired from their own telemedicine center in Riyadh, Saudi Arabia.[28] Distance education is also becoming an export business. Knowledge TV, a cable-television network based in Colorado and specializing in distance education, sells programming in Australia and Hong Kong. The University of Maryland University College, which specializes in part-time education, teaches courses at every military base in Europe and Asia, laying on computer conferences and voice-mail to allow students to communicate with their teachers and with one another. A growing number of universities offer courses over the Internet, including—understandably—courses about the Internet.

The Size of the Nation State

. . . . .

Not only will the size of government shrink, but so also may the size of nations. Indeed, that has been the trend throughout the second half of the twentieth century. The number of independent countries has almost doubled since 1960. Since the late 1980s, the collapse of the Soviet Union, the disintegration of Yugoslavia, and the division of Czechoslovakia have all added to the number of nation states. More nations will emerge, some peacefully, some violently, if separatist movements continue to gather strength.

Many of these new countries are small. In the Commonwealth, for instance, a club to which many of the century's newly independent nations belong, half the members have populations of less than one million—smaller than Portland, Oregon, or Marseilles, France. The economies of small countries are usually extremely open to international trade. Hong Kong, Singapore, and Luxembourg are merely the most prosperous examples of small countries that have thrived on the growth of global commerce. The export production of small countries is also highly specialized. In many of the Commonwealth's small states, five kinds of products account for about half of total exports.[29]

In the past, the main economic advantage of being part of a large country was to have access to a large marketplace. Small countries thus have a particular interest in anything that allows them to treat the entire world as their home market. They thus have a much greater interest than larger countries in a liberal international trade regime that allows them to export freely without the hindrance of tariff barriers. They also have a special need for inexpensive communications.

That, incidentally, applies to air fares just as much as to telecommunications. Almost every international air route is more expensive than a domestic route of similar length would be. Rather than struggling in vain to run national airlines, protecting them by maintaining fare cartels, many small countries would be far wiser to open their airports to any foreign carrier willing to undercut rivals.

Small countries—almost by definition—often occupy the fringes of large markets. Like small companies, they need to become global niche players, selling specialist products to the biggest market of all: the world market. The Internet allows them to advertise their wares worldwide; the telephone, to take and dispatch orders. Countries such as Ireland (population: 3.6 million) and Iceland (population: 260,000) have already spotted this new opportunity. Both use their well-educated work forces to foster communications-based businesses.

Low-cost communications will not, of course, remove all the economic handicaps that small countries suffer. Their greater specialization will still make them more vulnerable than are bigger nations to external shocks. But for independence movements, the communications revolution is good news. Not only can they fight their wars of liberation with the laptop and the Web site, but once they succeed, their new mini-nation has a better chance of survival than it would have had in the past. The combination of a liberal global trading regime and good communications would allow an independent Scotland or Quebec to compete on less unequal terms than ever before with the world's bigger countries.

Communications and Peace

. . . . .

In January 1815, the British and Americans fought the battle of New Orleans—unnecessarily: the war of 1812 in Europe had finished a fort-

. . . . .

night earlier, but the news had not yet reached North America.[30] Good communications between governments have always been the bedrock of peace: governments are more likely to fight one another if each does not have a clear idea of the intentions of the others.

By the early years of the new millennium, communications will cement the nations of the world together, at three levels. First, governments can be better informed than ever before about what other governments are doing. Direct links between governments have come a long way since the installation of the hot line between the Kremlin and the White House. In addition, diplomats who learn to use the Internet (still a rare breed) can trawl through material that once could have been extracted only by a skilled ambassador. You want the full text of the Hebron agreement, complete with appendices and detailed military maps? A few clicks of the mouse, and it can be on your screen.

While such legitimate inquiry is easier, so are new kinds of intrusion. Hacking into another country's computer systems is a great deal safer (if less romantic) than spying used to be. And in future, armies will be vulnerable to digital sabotage. The more important communications become to war and government, the more vulnerable they are to jamming, viruses, and interference. Defense strategy in the future will be as much about computer and communications skills as about building bigger bombs.

Second, the death of distance means that the countries of the world will be tied together by innumerable commercial bonds. Many more companies will have subsidiaries and operations on every continent, or at least in every big time zone. Countries that invest in one another are much less likely to fight one another.

Above all, ordinary citizens will learn more about people in other countries. At present, it costs more to call a foreign country, however close, than to call another part of one's own land, however distant. Few individuals use mail order or the telephone to shop abroad; even fewer watch foreign television channels or read foreign newspapers. Such activities will now be far easier and far less expensive.

People will now grow used to the idea of shopping abroad, not just for goods but also for services. Buy a foreign car, and you learn nothing about the country where it was made. Buy a foreign movie or CD or use a foreign secretarial service, and you start to build a relationship with another nation. As international trade in services builds up, it will create a firmer bond between countries than trade in goods has done. Such activities will help to shrink the world and to make people realize the extent to which, in John Donne's words, "No man is an island, entire of itself; every man is a piece of the continent, a part of the main."

To communicate may not be enough to keep the nations of the earth at peace with one another, but it is a start. Free to explore different points of view, on the Internet or on the thousands of television and radio channels that will eventually be available, people will become less susceptible to propaganda from politicians who seek to stir up conflicts. Bonded together by the invisible strands of global communications, humanity may find that peace and prosperity are fostered by the death of distance.

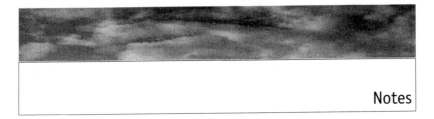

Notes

Preface

1. "The Death of Distance: A Survey of Global Telecommunications," *The Economist*, 30 September 1995. The survey, and all other *Economist* articles mentioned in this book, are available (on subscription) at www.economist.com. A further telecommunications survey by the author, "A connected world", appeared in *The Economist*, 13 September 1997.
2. Geoffrey Blainey, *The Tyranny of Distance* (Sydney: Pan Macmillan, 1966).
3. Simon Forge, "Near-Zero Tariff Telecommunications" (Cambridge Strategic Management Group, Paris, unpublished manuscript).
4. Hamish McRae, *The World in 2020: Power, Culture and Prosperity* (Boston: Harvard Business School Press, 1994).

Chapter 1

1. Data from Matrix Information and Directory Services (www.mids.org), January 1997.
2. Quoted in *Beyond the Internet: Restructuring the Communications Market*, ed. Sam Cover (Cambridge, England: Analysys Publications, 1996), ix.
3. Melvin Harris, *ITN Book of Firsts* (London: Michael O'Mara Books, 1994), 108, 124, 162.
4. John Pearson, *Painfully Rich: J. Paul Getty and His Heirs* (London: Macmillan, 1995), 68.
5. A cable from the mainland United States to Hawaii that went on-line the following year had only ninety-one such "voice paths." Gregory C. Staple, ed., *TeleGeography 1995: Global Telecommunications Traffic Statistics and Commentary* (Washington, D. C.: TeleGeography, 1995), 84.
6. Walter B. Wriston, *The Twilight of Sovereignty: How the Information Revolution Is Transforming our World* (New York: Scribner's, 1992), 36.
7. Gregory C. Staple, ed., TeleGeography 1996–97: Global Telecommunications Traffic Statistics and Commentary (Washington, D. C.: TeleGeography, 1996), xv.
8. International Telecommunication Union Database, Geneva, May 1997.

9. Stephen Goodwin, "Trapped on Everest? I'm on My Mobile," *Independent* 27 April 1996, 3. Eamon Fullen's rescue began, as newspapers reported at the time, with "the standard procedure for males in distress—call the wife." His wife took the call, made from his solar-powered satellite phone, two thousand miles away, in Hong Kong, and alerted the Nepalese army, which organized a helicopter rescue. Mobile telephones have become the bane of Scotland's mountain-rescue teams because they encourage hill-walkers to ignore the promptings of common sense.

10. World Telecommunication Development Report 1996/97 (Geneva: International Telecommunication Union, 1997), A-75, Table 18.

11. *Defining Moments* (London: A. T. Kearney, private publication, 1996), 46.

12. Robert H. Frank and Philip J. Cook, *The Winner-Take-All Society: How More and More Americans Compete for Ever Fewer and Bigger Prizes, Encouraging Economic Waste, Income Inequality, and an Impoverished Cultural Life* (New York: The Free Press, 1995).

13. Quoted in Victor Navasky, "Tomorrow Never Knows," *New York Times Magazine,* 29 September 1996, 216.

14. *Defining Moments,* 38.

15. Chris Anderson, personal communication, February 1997.

16. "Computer Power and Data Storage Capacity," *Screen Digest,* December 1996, 287.

17. Chris Anderson, "The Accidental Superhighway: A Survey of the Internet," *The Economist,* 1 July 1995.

18. Patrick Butler, Ted W. Hall, Alistair M. Hanna, Lenny Mendonca, Byron Auguste, James Manyika, and Anupam Sahay, "A Revolution in Interaction," *McKinsey Quarterly* 1 (1997): 7. The McKinsey team argues that "interactions" represent 51 percent of labor activity in the United States and 46 percent in Germany.

19. Quoted by Robert Pepper in "Towards a Market-Based Spectrum Policy" (speech given at Spectrum '96, London, May 1996).

20. *Principles of Economics,* 8th ed. (1920, book IV). Quoted in *Beyond the Internet,* ed. Sam Cover (Cambridge, England: Analysys Publications, 1996), 1.

Chapter 2

1. International Telecommunication Union, *World Telecommunication Development Report 1995* (Geneva: International Telecommunication Union, 1995), 3.

2. Price Waterhouse, *Technology Forecast: Entertainment, Media and Communications* (Menlo Park, Calif., Price Waterhouse, 1995), 109.

3. Gregory C. Staple, ed., *TeleGeography 1995: Global Telecommunications Traffic Statistics and Commentary* (Washington, D. C.: TeleGeography, 1995), 86–87.

4. International Telecommunication Union and TeleGeography, Inc., *Direction of Traffic: Trends in International Telephone Tariffs* (Geneva and Washington: International Telecommunication Union and TeleGeography, 1996), 5.

5. "Undersea Fibre: From Toll Road to Trading Pit," *Public Network Europe* 5, no. 7 (July/August 1995): 31–34.

6. One of the biggest of these private networks connects the computer terminals of the world's airlines and travel agencies: SITA, or the Société Internationale de Télécommunications Aéronautiques. Another links the world's banks: SWIFT, or the Society for Worldwide Interbank Financial Telecommunications.

7. International Telecommunication Union and TeleGeography, *Direction of Traffic*, 69.

8. Ibid., 19.

9. Jennifer Schenker, "USA Global Link Today Will Launch World-Wide Internet Phone Service," *Wall Street Journal Europe*, 24 March 1997.

10. Quoted in Frances Cairncross, "The Death of Distance: A Survey of Telecommunications," *The Economist*, 30 September 1995, 10.

11. Zachary M. Schrag, "The Internet as a Telephone Service: No Free Lunch," in *TeleGeography 1995: Global Telecommunications Traffic Statistics and Commentary*, ed. Gregory C. Staple (Washington, D. C.: TeleGeography, 1995), 59.

12. International Telecommunication Union and TeleGeography, *Direction of Traffic*, 51.

13. Ibid., 62.

14. Susanna Cresswell, "The Shock of Recognition," *Voice International*, December 1996/January 1997, 32–36.

15. Kim Fennel, "Messaging into the Millenium," *Voice International*, December 1996/January 1997, 24–25.

16. Gregory C. Staple, ed., *TeleGeography 1996–97: Global Telecommunications Traffic Statistics and Commentary* (Washington, D. C.: TeleGeography, 1997), xiii.

17. Estimate as of 1 January 1997 by the International Telecommunication Union.

18. *Mobile and PSTN Communication Services: Competition or Complementarity?* (Paris: OECD, 1995). Available at www.oecd.org/dsti/gd.docs/gdlist_e.html.

19. International Telecommunication Union, *African Telecommunication Indicators 1996* (Geneva: International Telecommunication Union, 1996).

20. Peter W. Huber, Michael K. Kellogg, and John Thorne, *The Geodesic Network II: 1993 Report on Competition in the Telephone Industry* (Washington D. C.: The Geodesic Company, 1992).

21. John J. Keller, "AT&T to Test Linking Homes to Its Wireless Network," *Wall Street Journal Europe*, 25 February 1997.

22. Vanessa Houlder, "On the Right Wavelength," *Financial Times*, 21 January 1997.

23. Study by the Internet Access Coalition, quoted by Louise Kehoe, "Home Telephones under Siege," *Financial Times*, 2 February 1997.

24. "Surfing the 'Second Wave': Sustainable Internet Growth and Public Policy." Available at http://www.pactel.com/about/pub_policy/esp/WP_internet-part1.html.

25. Nicholas Negroponte, *Being Digital* (London: Hodder & Stoughton, 1995), 26.

26. Robert Allen (speech before the Council on Foreign Relations, Washington D. C., 26 October 1994).

27. Patrick Butler, Ted W. Hall, Alistair M. Hanna, Lenny Mendonca, Byron Auguste, James Manyika, and Anupam Sahay, "A Revolution in Interaction," *McKinsey Quarterly* 1 (1997): 12.

28. International Telecommunication Union, *World Telecommunication Development Report 1996/97* (Geneva: International Telecommunication Union, 1997), A-15.
29. International Telecommunication Union, *World Telecommunication Development Report 1994*, 87.
30. Ibid., 75.

Chapter 3

1. Bill Bryson, *Made in America: An Informal History of the English Language in the United States* (New York: William Morrow, 1995).
2. A survey by Emerging Technologies Research Group of New York found that 32 percent of the people who get Internet access watch less television than before; cited in Frank Rose, "The End of TV as We Know It," *Fortune*, 23 December 1996, 62.
3. Zaire's broadcasting authority used to begin its evening news programs with a sequence showing President Mobutu Sese Seko descending to earth from the clouds.
4. The average capacity of European cable systems is twenty-seven channels; cited in "Europe's 'Other' Channels," *Screen Digest*, March 1997, 57.
5. Richard Kee, John Davison, Mari Vahanissi, and Kate Hewett, *Cable: The Emerging Force in Telecoms and Interactive Markets* (London: Ovum, 1996).
6. Ibid.
7. International Telecommunication Union, *World Telecommunication Development Report 1995* (Geneva: International Telecommunication Union, 1995), 58.
8. International Telecommunication Union, *World Telecommunication Development Report 1994* (Geneva: International Telecommunication Union, 1994), 33.
9. "Digital Television: Start of the Worldwide Lift-Off," *Screen Digest*, August 1996, 177.
10. Price Waterhouse, *Technology Forecast: Entertainment, Media, and Communications* (Menlo Park, Calif., Price Waterhouse 1995),157.
11. Paul Farhi, "NBC Scores Serious Coup by Salvaging Hit Comedy," *International Herald Tribune*, 14 May 1997.
12. Steven Heyer, Turner Broadcasting System, telephone conversation with author, April 1997.
13. For instance, British children aged 2 to 9 in homes with cable or satellite devoted 64 percent of their viewing to cable or satellite channels in October 1995; the share for adults in such homes was 36 percent. John Clemens and Jane Key, *Trends in Viewing in Cable TV Homes 1990–95* (London: Independent Television Commission, 1996), 7.
14. Between 1990 and 1994, both the prime-time ratings and the audience share of the networks held steady, although they dipped in 1995. Veronis, Suhler and Associates, *Communications Industry Forecast* (New York: Veronis, Suhler and Associates, 1996), 45.
15. In the United States, where prime-time television programs broadcast by the three older networks typically reach ten to twelve million households apiece, an advertiser pays between 1.2 and 1.4 cents per household for a thirty-second slot ("Liberty Media Group," Schroder Wertheim, New York, September 1996).

16. Veronis, Suhler and Associates, *Communications Industry Forecast*, 53.

17. "Murdoch's Empire," *The Economist*, 9 March 1996, 101.

18. Rebecca Winnington-Ingram, ". . . And All Shall Have Prizes . . .?" (speech before Morgan Stanley conference, London, 1996), slide 2.

19. In Britain, between 1990 and 1995, the amount of time spent viewing television of all kinds actually fell, even in homes with cable, and the overall share of time spent watching cable and satellite as opposed to broadcast channels did not increase. See John Clemens and Jane Key, *Trends in Viewing in Cable TV Homes 1990–96* (London: Independent Television Commission, 1996), 5.

20. Veronis, Suhler and Associates, *Communications Industry Forecast*, 26.

21. Japan is developing a concept called "refrigerator television," through which television programs will be recorded automatically in the television set's built-in memory (the amount of which will be increased over time, as PC memory has been). The viewer will then be able to "reheat" previously broadcast material at will. Tim Kelly, International Telecommunication Union, conversation with author, February 1997.

22. "Consuming Movies: Pay TV Eats into Film Spending Cache," *Screen Digest*, January 1997, 9.

23. International Telecommunication Union, *World Telecommunication Development Report 1996–97*, A-75.

24. International Telecommunication Union, *World Telecommunication Development Report 1995*.

25. Price Waterhouse, *Technology Forecast: Entertainment, Media and Communications* (Menlo Park, Calif., Price Waterhouse, 1995), 109.

26. John Malone (comments at Technology Day, Telecommunications Inc., 11 January 1996, unpublished transcript), 12.

27. Tim Jackson, "WebTV Waits for Green Light," *Financial Times*, 24 March 1997.

28. Tom Wolzein and John Penny, "Danger in Videoland: Is a Commodity Revenue Spiral Ahead?" (circular from Sanford Bernstein, New York, 1996).

29. "The Tangled Webs They Weave," *The Economist*, 16 October 1993.

30. Mike France, "The NBA vs AOL: You Gotta Pay to Play," *Business Week*, 16 September 1996.

Chapter 4

1. Matrix Information and Directory Services, www.minds.org, January 1997.

2. The survey method used by MIDS, the source of the figures in Figure 4-2, is different and gives a higher result than that used by Network Wizards, the other main source of such data.

3. In one sign of its maturity, the Internet now has its own historians, including Katie Hafner and Matthew Lyon, whose account of the pioneers, in *Where Wizards Stay up Late* (New York: Simon & Schuster, 1996), provides the best account of the Internet's early days.

4. The first—temporary—link beyond American shores was to Brighton, England, in 1973 for a conference on computing.

5. In a further sign of American dominance, American domain names rarely carry a country tag. The domain tag ".us" exists, but it is not much used. On a similar principle, only Britain, the country where the postage stamp was invented, does not place its name on its stamps.
6. "Nets without Frontiers," *Digital Media 5*, vol. 10 (12 March 1996): 19.
7. Sam Paltridge, "How Competition Helps the Internet," *OECD Observer* no. 201 (August/September 1996): 25.
8. Survey by Find/SVP and Jupiter Communications, October 1996, available at www.jup.com/jupiter/release/oct96/.
9. Sam Paltridge, *Information Infrastructure Convergence and Pricing: The Internet* (Paris: OECD, 1996), 30.
10. Colin Blackman and Michael Denmead, *1998: A New Era for EU Telecoms Regulation*, ed. Kate Gentles (Cambridge, England: Analysys, 1996).
11. Paltridge, "How Competition Helps," 27.
12. OECD, *Communications Outlook 1997* (Paris: OECD, 1997), 115.
13. Bill Gates, *The Road Ahead* (New York: Viking Penguin, 1995), 91.
14. Chris Anderson, "A World Gone Soft: A Survey of the Software Industry," *The Economist*, 25 May 1996, 20.
15. Chris Anderson, "The Accidental Superhighway: A Survey of the Internet," *The Economist*, 1 July 1995, 9.
16. Richard Behar, "Who's Reading Your E-Mail?" *Fortune*, 3 February 1997, 31.
17. Goldman Sachs, quoted in a report for the British Department of Trade and Industry prepared by Spectrum Strategy Consultants, Development of the Information Society (Norwich: HMSO, 1996), 90.
18. J. William Gurley, "E-Mail Gets Rich," *Fortune*, 17 February 1997, 67–68.
19. Vanessa Houlder, "Failing to Get the Message," *Financial Times*, 17 March 1997.
20. Amy Cortese, "A Way Out of the Web Maze," *Business Week*, 27 February 1997.
21. "The Total Librarian," *The Economist*, 14 September 1996.
22. Cortese, "A Way Out."
23. Louise Kehoe, "Big Rise in Hacker Break-Ins," Information Technology Review, *Financial Times*, 3 July 1996.
24. Quoted in Behar, "Who's Reading?" 30.
25. Kehoe, "Big Rise."
26. Quoted in Cortese, "A Way Out."
27. "The Interminablenet," *The Economist*, 3 February 1996.
28. "The War of the Waves," *The Economist*, 11 May 1996.
29. "Too Cheap to Meter?" *The Economist*, 19 October 1996.
30. The Internet Society, telephone conversation with author, 28 April 1997.
31. Anderson, "Accidental Superhighway," 13.
32. Many of the new businesses started in Britain in 1996, a record year for start-ups, had the word "net" in their names.

Chapter 5

1. According to Nicholas Denton ("Drive to Plug the Gap," *Financial Times*, 3 February 1997), venture capital accounts constitute 0.8 percent of GDP in the United States—and in 1995, 78 percent of venture capital investments were in high-tech companies.
2. Richard S. Tedlow, "Roadkill on the Information Superhighway," *Harvard Business Review*, November/December 1996, 15.
3. Mary Meeker, *The Internet Advertising Report*, Morgan Stanley U.S. Investment Research, December 1996.
4. Chris Anderson, "In Search of the Perfect Market: A Survey of Electronic Commerce," *The Economist*, 10 May 1997.
5. Quoted in "This Section of the Global Information Infrastructure Is Sponsored by . . .," International Telecommunication Union, *World Telecommunication Development Report 1995* (Geneva: International Telecommunication Union, 1995), 29.
6. Nicholas Denton and Hugh Carnegy, "Ad Breaks Are the Price for Free Swedish Phone Calls," *Financial Times*, 20 January 1997.
7. Anderson, "In Search."
8. John Hagel III and Arthur G. Armstrong, *Net Gain: Expanding Markets through Virtual Communities* (Cambridge, Mass.: Harvard Business School Press, 1997), 192–193.
9. "Electronic Commerce Comes of Age" (AT&T press release, 14 March 1997).
10. International Telecommunication Union, *World Telecommunication Development Report 1995*, 96.
11. "Electronic Retailing: Interactive Potential," *Screen Digest*, June 1996, 133.
12. "Home Alone?" *The Economist*, 12 October 1996.
13. Tedlow, "Roadkill," 15.
14. "The Interactive Home," INTECO Research, Norwalk, Conn., 1994.
15. "Suited, Surfing, and Shopping," *The Economist*, 25 January 1997.
16. Alexa Kierzkowski, Shayne McQuade, Robert Waitman, and Michael Zeisser, "Marketing to the Digital Consumer," *McKinsey Quarterly* 3 (1996): 9–10.
17. Patrick Butler, Ted W. Hall, Alistair M. Hanna, Lenny Mendonca, Byron Auguste, James Manyika, and Anupam Sahay, "A Revolution in Interaction," *McKinsey Quarterly* 1 (1997): 19.
18. Steve Homer, "Fisher Offers Sea of Possibilities in the World of Heavy Industry," *The Independent*, 22 April 1996.
19. Price Waterhouse, *Technology Forecast: Entertainment, Media and Communications* (Menlo Park, Calif.: Price Waterhouse, 1997), 445–446.
20. Anderson, "In Search."
21. Peter Norman, "German Companies Blame Internet for Export Decline," *Financial Times*, 27 March 1996.
22. "Going, going . . .," *The Economist*, 31 May 1997.
23. Perry Flint, "Cyber Hope or Cyber Hype?" *Air Transport World*, October 1996, 25–26.
24. Bill Gates, *The Road Ahead* (New York: Viking Penguin, 1995), 157.

· · · · ·

25. Amon Cohen, "Rescue in Cyberspace," *Financial Times*, 29 July 1996.

26. "Death of a Salesman," *The Economist*, 23 September 1995.

27. Fidelity's site is at personal.fidelity.com/decisions/college/calculator.html.

28. Jerrold M. Grochow, "Don't Bank on the Internet, Quite Yet," *Financial Times*, 16 April 1996.

29. "Brokers and the Web," *The Forrester Report* 2, no. 1 (September 1996).

30. Anderson, "In Search."

31. Katie Hafner, "Log On and Shoot," *Newsweek*, 12 August 1996.

32. Iver Peterson, "Wall Street Journal on Line: Readers Pay but Profits Remain Elusive," *The New York Times*, 10 February 1997.

33. Gregory C. Staple, ed., "Settlements for Phone Sex," *TeleGeography 1996/97: Global Telecommunications Traffic Statistics and Commentary* (Washington, D.C.: TeleGeography, 1996).

34. Harold L. Vogel, *Entertainment Industry Economics* (Cambridge: Cambridge University Press, 1994), 222.

35. Legitimate gambling can be high-tech, too: the computer network of Britain's National Lottery, which links the service tills in shops all over the country, is larger than that of Britain's four main high-street banks put together.

36. Evan I. Schwartz, "Wanna Bet?" *Wired*, October 1995, 137.

37. USA Today/IntelliQuest survey, quoted by The Boston Consulting Group in *The Information Superhighway and Retail Banking*, (Boston, 1995).

38. Quoted in Anderson, "In Search."

39. "Enter the Intranet," *The Economist*, 13 January 1996.

40. Peter Martin, "The Courage to Open Up," *Financial Times*, 31 October 1996.

41. Quoted by Louise Kehoe in "Cultural Chasm," *Financial Times*, 19 February 1997.

42. Geoff Nairn, "Awaiting the Virtual Call," *Financial Times*, 13 June 1996.

43. Butler et al., "A Revolution in Interaction," 16.

44. "GE Launches New Internet-Based Commerce Network, Free Supplier Training Scheduled," PRN Newswire, on-line, 2 January 1996.

45. Anderson, "In Search," 12–13.

46. Diane Summers, "Bespoke Jeans for the Masses," *Financial Times*, 12 September 1996.

47. Erik Brynjolfsson, Thomas Malone, Vijay Gurbaxani, and Ajit Kambil, "Does Information Technology Lead to Smaller Firms?" *Management Science* 40, no. 12 (December 1994).

48. Keith Bradley, "The Value of Intellectual Capital," *Financial Times*, 26 July 1996.

Chapter 6

1. Reed Hundt (speech before the "Beyond the Telecom Act" Conference, Freedom Forum, Arlington, Va., 7 February 1997).

2. See, for instance, Jerry Hausman and Timothy Tardiff, "Valuation and Regulation of New Services in Telecommunications," in *The Economics of the Information*

Society, ed. A. Dumort and J. Dryden (Luxembourg: Office for Official Publications of the European Communities, 1997).

3. David Crystal, *Cambridge Encyclopedia of the English Language,* (Cambridge, England: Cambridge University Press, 1995), 108.
4. Jube Shiver, Jr., "America Takes a Step Toward Digital TV," *International Herald Tribune,* 27 November 1996.
5. "Thoroughly Modern Monopoly," *The Economist,* 8 July 1995.
6. Stan Liebowitz and Stephen Margolis, "Network Externality: An Uncommon Tragedy," *Journal of Economic Perspectives* 8, no. 4 (Spring 1994): 133–150.
7. Telecoms Regulation in Europe, EIU Report No. P512 (London: The Economist Intelligence Unit, 1995), 55.
8. Ibid.
9. Between 1988 and 1993, countries that had only one monopoly international operator saw minutes of international call traffic per main line grow by nearly 13 percent; in the United States and Japan, the two industrial countries with more than two international carriers, growth averaged over 20 percent. International Telecommunication Union, *World Telecommunication Development Report 1995* (Geneva: International Telecommunication Union, 1995), 114–116.
10. The shift from analogue to digital cellular systems created a second chance for some countries that had originally licensed only one carrier to create competition. Several European countries, such as Germany, Spain, and Portugal, had few analogue subscribers but introduced competition when Europe developed a digital standard.
11. Note too, another finding: the average prices charged by an Internet access provider for "dial-up" services—where a customer with a PC and a modem simply dials in using an ordinary telephone line—are on average nearly three times lower (and sometimes much more) in competitive markets such as Australia and New Zealand than in monopoly ones such as Germany and Spain.
12. To put that in perspective, the cost of a typical basket of ordinary telephone services for residential users in the most expensive countries costs only two to three times more than in the least expensive countries.
13. Michael Richardson, "Slowly, Asian Giants Open Door to Competition," *International Herald Tribune,* 13 March 1997.
14. The obligation did not seem to deter license applicants; ten companies were awarded licenses, with an obligation to install 4.4 million telephone lines by 1999. The national carrier, faced with real competition for the first time, set up a scheme, called the Zero Backlog Program, to wipe out its long waiting list. The installation of new lines, once slouching at around 20,000 per year since the mid-1980s, shot up to 200,000 in 1993 and 250,000 in 1994. Source: International Telecommunication Union, *World Telecommunication Development Report 1995,* 114.
15. Frances Cairncross, "The Death of Distance: A Survey of Global Telecommunications," *The Economist,* 30 September 1995, 26.
16. Ibid.

17. One scheme of this sort is described in Telecoms Regulation in Europe, Research Report No. P512 (London: The Economist Intelligence Unit, 1995).
18. W. Brian Arthur, "Increasing Returns and the Two Worlds of Business," *Harvard Business Review,* July/August 1996.

Chapter 7

1. Christopher Walker, "Orthodox Jews Go Surfing on the Kosher Internet," *Times* (London), 21 October 1996.
2. Ithiel de Sola Pool, *Technologies of Freedom* (Cambridge, Mass: Belknap Press, 1983).
3. Quoted in "The Coming Global Tongue," *The Economist,* 21 December 1996.
4. James Mackintosh, "Internet Access Provider Boosts Efforts to Censor Pornography," *Financial Times,* 6 May 1996.
5. Michael Meyer, "Whose Internet Is It?" *Newsweek,* 22 April 1996.
6. "Heavy Breathing," *The Economist,* 30 July 1994.
7. "U.S. Court Overturns Law to Curb Internet," *International Herald Tribune,* 13 June 1996.
8. Meyer, "Whose Internet Is It?"
9. James Kynge, "Electronic Undesirables," *Financial Times,* 9 September 1996.
10. "The Top Shelf," *The Economist,* 18 May 1996.
11. Ibid.
12. Gretchen Atwood, "Going Beyond Blocking," *Digital Media* 5, no. 11 (8 April 1996): 3.
13. Marie D'Amico, "Court Rules CDA Unconstitutional," *Digital Media* 6, no. 1 (June 1996): 5.
14. Silvia Ascarelli and Kimberley A. Strassel, "German Cases Illuminate Struggle to Regulate Net," *Wall Street Journal Europe,* 21 April 1997.
15. "We Know You're Reading This," *The Economist,* 10 February 1996.
16. Simon Hughes and Paul Thompson, "Mobile Phones Left Trail for Cops," *The Sun,* 5 March 1997.
17. Phil Reeves, "Rumours Run Wild about Dead Leader," *The Independent,* 30 April 1996.
18. The legislation was passed after a newspaper published the video-rental list of Clarence Thomas, a newly nominated Supreme Court judge.
19. International Telecommunication Union, *World Telecommunication Development Report 1995* (Geneva: International Telecommunication Union, 1996), 122.
20. H. Stuart Taylor, letter, "If the Dates Fit, Then Capitalise on Them," *Financial Times,* 1 August 1996.
21. Peter Huber, *Orwell's Revenge* (New York: The Free Press, 1994).
22. Alexandra Wyke, "The Future of Medicine," in *Going Digital: How New Technology Is Changing Our Lives* (London: *The Economist* in association with Profile Books, 1996).
23. "Online Prying Made Easy," *Business Week,* 30 September 1996.

24. Sverker Lindbo, e-mail to author, April 1997.

25. Margie Wylie, "Free E-Mail," *Digital Media* 5, no. 12 (14 May 1996): 3.

26. Michael Cassell, "Freeze on Cold Calls," *Financial Times*, 8 August 1996.

27. This question is explored further by John Hagel III and Jeffrey F. Rayport in "The Coming Battle for Customer Information," *Harvard Business Review*, January/February 1997.

28. "You Name It" (letter), *The Economist*, 13 July 1996.

29. "Names Writ in Water," *The Economist*, 8 June 1996.

30. Jennifer L. Schenker and Rebecca Quick, "Debate over the Right to Assign Addresses Shakes Up the Internet," *Wall Street Journal Europe*, 30 April 1997.

31. Nikki Tait, "Australians in a Spin over Cost of Music," *Financial Times*, 21 March 1997.

32. de Sola Pool, *Technologies of Freedom*, 249.

Chapter 8

1. Pam Woodall, "A Hitchhiker's Guide to Cybernomics: A Survey of the World Economy," *The Economist*, 28 September 1996, 43.

2. Ibid.

3. "Accountants Want to Be Faceless," *Net Profit*, June 1997, 7. www.net-profit.co.uk.

4. Alexandra Wyke, "The Future of Medicine," in *Going Digital: How New Technology Is Changing Our Lives* (London: *The Economist* in association with Profile Books, 1996), 242.

5. *Sunday Telegraph*, 17 November 1996.

6. Woodall, "Hitchhiker's Guide," 43.

7. Ithiel de Sola Pool, "The Communications/Transportation Tradeoff," *Current Issues in Transportation Policy*, ed. Alan Altschuler (Lexington, Mass.: D.C. Heath, 1979), 182.

8. Robert J. Saunders, Jeremy J. Warford, Björn Wellenius, *Telecommunications and Economic Development*, 2nd ed., International Bank for Reconstruction and Development (Baltimore, Md.: Johns Hopkins University Press, 1994), 130.

9. Ibid.

10. "Bangalore Bytes," *The Economist*, 23 March 1996.

11. Shailagh Murray, "Rise of Tele-Business in Ireland Gives Jobless in EU New Prospects," *Wall Street Journal Europe*, 5 February 1997.

12. Andrew Bibby, "Hebrides Telecommuters Fight Isolation with High Technology," *International Herald Tribune*, 24 March 1997.

13. Alan Cane, "New Lines of Attack," *Financial Times*, 12 September 1996.

14. Quoted in Shailagh Murray, "Rise of Tele-Business in Ireland Gives Jobless in EU New Prospects," *Wall Street Journal Europe*, 5 February 1997.

15. John Heilemann, "It's the New Economy, Stupid," *Wired*, March 1996, 70.

16. Stephen Golub, Comparative and Absolute Advantage in the Asia-Pacific Region, Federal Reserve Bank of San Francisco working paper, 1995.

17. See Robert H. Frank and Philip J. Cook, *The Winner-Take-All Society: How More and More Americans Compete for Ever Fewer and Bigger Prizes, Encouraging Economic Waste, Income Inequality, and an Impoverished Cultural Life* (New York: The Free Press, 1995). The idea was first described by Sherwin Rosen in "The Economics of Superstars" *American Economic Review* 71 (December 1981): 845–858.
18. Frank and Cook, *Winner-Take-All Society*, 3.
19. Woodall, "Hitchhiker's Guide."
20. "Computers Put the Zip in the GDP," *Business Week*, 4 November 1996, 206.
21. Quoted in Woodall, "Hitchhiker's Guide," 13.
22. Ibid., 15.
23. Ibid., 13.
24. Paul David, "The Dynamo and the Computer: An Historical Perspective on the Modern Productivity Paradox," *American Economic Review*, May 1990.
25. Patrick Butler, Ted W. Hall, Alistair M. Hanna, Lenny Mendonca, Byron Auguste, James Manyika, and Anupam Sahay, "A Revolution in Interaction," *McKinsey Quarterly* 1 (1997): 10.
26. Woodall, "Hitchhiker's Guide," 15.

Chapter 9

1. During BT's one-year experiment, eleven operators worked from home, monitored by psychologists from Aberdeen University, and handled 750,000 customer inquiries. BT home page, www.labs.bt.com/library/on-line/tele-work/.
2. Carl Frankel, "The Telecommuting Edge," *Tomorrow* 7, no. 1 (January/February 1997).
3. Charles Handy, *The Age of Unreason* (Boston: Harvard Business School Press, 1990).
4. See William Gibson, *Neuromancer* (New York: Ace Books, 1984).
5. International Telecommunication Union, *World Telecommunication Development Report 1994* (Geneva: International Telecommunication Union, 1994), 14.
6. John Hagel III, Ennius E. Bergsma, and Sanjeev Dheer, "Placing Your Bets on Electronic Networks," *McKinsey Quarterly* 2 (1996): 59.
7. Howard Rheingold, *The Virtual Community: Homesteading on the Electronic Frontier* (Reading, Mass.: Addison-Wesley, 1993).
8. Neil MacLean, "Clans," *High Life*, September 1996. See also the Electronic Scotland Web site.
9. Marc Peyser and Claudia Kalb, "Roots Network," *Newsweek*, 3 March 1997.
10. Quoted by David Crystal in *Cambridge Encyclopedia of the English Language* (Cambridge, England: Cambridge University Press, 1995), 386.
11. Crystal, *Cambridge Encyclopedia of the English Language*, 106.
12. Geoffrey Nunberg, quoted in "The Coming Global Tongue," *The Economist*, 21 December 1996.
13. Robert H. Frank and Philip J. Cook, *The Winner-Take-All Society: How More and More Americans Compete for Ever Fewer and Bigger Prizes, Encouraging Economic*

Waste, Income Inequality, and an Impoverished Cultural Life (New York: The Free Press, 1995).

14. Study by David Crystal, quoted in "The Coming Global Tongue," *The Economist,* 21 December 1996.
15. Christopher Johnston, "In Wales, TV Helps a Language Live," *International Herald Tribune,* 27 November 1996.
16. Nunberg, quoted in "The Coming Global Tongue."
17. "Gumped," *The Economist,* 24 December 1994.
18. Martin Dale, *The Movie Game* (London: Cassell, 1997).
19. "European Producers Start to Make Programs Europeans Want to Watch," *TV International,* 18 November 1996.
20. Victoria Griffiths, "Strategy for Block Release," *Financial Times,* 30 September 1996.
21. International Telecommunication Union, *World Telecommunication Development Report 1994* (Geneva: International Telecommunication Union, 1994), 80.
22. International Telecommunication Union, *World Telecommunication Development Report 1995* (Geneva: International Telecommunication Union, 1995), 137.
23. Ibid.

Chapter 10

1. "Why the Net Should Grow Up," *The Economist,* 19 October 1996.
2. Eli M. Noam, "Electronics and the Dim Future of the University," *Science,* 13 October 1995, 247–249.
3. Chris Hedges, "Serbs Discover an Instrument of Revolt: Internet," *International Herald Tribune,* 9 December 1996.
4. Sverker Lindbo, e-mail to author, April 1997.
5. Jonathan Rauch, *Demosclerosis* (New York: Times Books, 1994).
6. James S. Fishkin, *Democracy and Deliberation* (New Haven, Conn.: Yale University Press, 1991), 67.
7. Alvin and Heidi Toffler, *Creating a New Civilization* (Atlanta: Turner Publishing, 1995).
8. "E-lectioneering," *The Economist,* 17 June 1995.
9. David Butler and Austin Ranney, eds., *Referendums Around the World: The Growing Use of Direct Democracy* (New York: Macmillan, 1994), 5.
10. Brian Beedham, "Full Democracy," *The Economist,* 21 December 1996.
11. F. Christopher Arterton, *Teledemocracy: Can Technology Protect Democracy?* Sage Library of Social Research Vol. 165 (Newbury Park, Calif.: Sage Publications, 1987).
12. United States, Department of the Treasury, Office of Tax Policy, Selected Tax Policy Implications of Electronic Commerce (Washington, D. C.: GPO, November 1996). Available on-line at www.ustreas.gov.
13. Jennifer Schenker, "OECD Countries Seek to Levy Sales Duties on Internet Commerce," *Wall Street Journal Europe,* 24 February 1997.
14. "Taxed in Cyberspace," *The Economist,* 13 July 1996.

15. Lisa Nishimoto, "Internet Sales Raise Tax Flag," *InfoWorld* 12 (August 1996).

16. "Digit-al Cash," *The Economist*, 15 June 1996.

17. "Bar-Coding the Poor," *The Economist*, 25 January 1997.

18. Heather E. Hudson, "Applications of Telecommunications for the Delivery of Social Services," *Telecommunications and Economic Development*, eds. Robert J. Saunders, Jeremy J. Warford, and Björn Welenius, International Bank for Reconstruction and Development (Baltimore: Johns Hopkins University Press, 1994), 344.

19. "Big Sister Is Watching You," *The Economist*, 11 January 1997.

20. Milt Freudenheim, "Video Approach to Hospital Care," *International Herald Tribune*, 27 February 1997.

21. Alexandra Wyke, "The Future of Medicine," in *Going Digital: How New Technology Is Changing Our Lives* (London: The Economist in association with Profile Books, 1996), 250.

22. Heather E. Hudson, "Applications of Telecommunications for the Delivery of Social Services," 349.

23. Noam, "Electronics and the Dim Future," 247–249.

24. David Brin, "The Transparent Society," *Wired*, December 1996, 62.

25. Michael Nelson, FCC, personal communication with author, 2 May 1997.

26. Scheherazade Daneshkhu and George Parker, "Ticketless Airline Travel Set to Become More Widespread," *Financial Times*, 23 October 1996.

27. Jim Flyznik, U.S. Treasury and head of Al Gore's technology team, telephone conversation with author, 19 May 1997.

28. "The Doctor Will See You Now—Just Not in Person," *Business Week*, 3 October 1994, 117.

29. Great Britain, Commonwealth Secretariat, *Small States: Economic Review and Basic Statistics* (London: HMSO, May 1996), 38–39.

30. Robert E. Allen, "The Borderless Superpower: Information Technology's Emerging Role in World Politics, Business, and Economic Growth" (speech before the U. S. Council on Foreign Relations, 26 October 1994).

Index

. . . . .